MAINTENANCE WORK MANAGEMENT

MAINTENANCE WORK MANAGEMENT

Design, Implement, and Sustain the Six MWM Processes

Dale R. Tanner

INDUSTRIAL PRESS, INC.

Industrial Press, Inc.
1 Chestnut Street
South Norwalk, Connecticut 06854
Phone: 203-956-5593
Toll-Free in USA: 888-528-7852
Email: info@industrialpress.com

Author: Dale R. Tanner
Title: Maintenance Work Management: Design, Implement, and Sustain the Six MWM Processes
Library of Congress Control Number: 2026946670

Printed in the United States of America.

ISBN (print): 978-0-8311-3707-6
ISBN (ePUB): 978-0-8311-9695-0
ISBN (eMOBI): 978-0-8311-9696-7
ISBN (ePDF): 978-0-8311-9694-3

Publisher/Editorial Director: Judy Bass
Compositor: Patricia Wallenburg, TypeWriting
Proofreader: David Johnstone
Copyeditor: Judy Duguid
Cover Designer: Jeff Weeks

books.industrialpress.com
ebooks.industrialpress.com
1 2 3 4 5 6 7 8 9 10

This book is dedicated to Denny Delsanter.
My mentor, my my confidant, my advisor, and most of all,
the best that friendship has to offer.

CONTENTS

PREFACE

First, let me define *maintenance work management.* Maintenance work management (MWM) is the structured process of identifying, prioritizing, planning, scheduling, executing, documenting, closing, and analyzing maintenance tasks such as repairs, inspections, projects, upgrades, and preventive care in order to optimize equipment reliability and minimize downtime.

Maintenance work management is a practical, easy-to-use roadmap for developing, using, and managing industrial maintenance work management processes that are embraced by all stakeholders in your organization. MWM will manage and sustain your behaviors with the use of understandable, measurable key performance indicators and their proven, quantifiable results. It's a step-by-step guide for developing your MWM coaches, analyzing your current processes, and designing, implementing, and sustaining the six maintenance work management processes:

- Work identification
- Work approval
- Work planning
- Work scheduling
- Work execution
- Work close out

These six processes will enable your transformation with a real project.

With two decades of designing and implementing operations work management processes as a supervisor, manager, director, general manager, and independent consultant, followed by over two decades of experience designing and implementing maintenance work management processes, I wanted to write this book to help others understand the many details of maintenance work management in an easy-to-understand format, I find great pleasure in hoping that this book helps explain how to identify and then gain the benefits from the science of maintenance work management.

INTRODUCTION

Maintenance work management is a science. It's not a personal preference for the untrained. I learned this while becoming a Certified Production and Inventory Manager (CPIM) with the American Production and Inventory Control Society (APICS), first studying books on production control and then earning an MBA in operations management. This was confirmed when working at my second refinery and chemical plant site when I learned that the client's maintenance work management standards were based on the APICS teachings.

On that project, I had the pleasure of learning from consulting professionals who each had been in maintenance work management for over 30 years after holding positions in manufacturing management. These very senior mentors taught me that the best results happen with boots in the dirt alongside the people who make the physical work happen. They taught me that the best path to maintenance work management process analysis, design, implementation, and sustainability is in collaboration with the frontline operations, maintenance, supply chain, and reliability people who stand at the workface. The workface is the location where the workers stand to perform the work every day, whose job it is to keep the machinery running as the engineers intended it to. This is the foundation of reliability and maximum production.

Early, at the start of my career, I learned by operating industrial machinery that maintenance is an investment, not a cost, while achieving my production goals every day. This wasn't hard, as I already had my share of experience operating performance and sports cars at a very young age. My investment in my equipment, to this day, can be seen in the smile on my face as I back into my garage.

This book is for the frontline stakeholders who show up every day to keep the means of production profitable in the most organized and cost-effective manner possible, and for the managers who can stand at the workface and know that it's their goal to support the people who get dirty and make it happen.

Throughout this book, you will see graphic examples of my work over the years. These are not intended to be used as standard templates, as each organization has its own personality, management, and culture. They do contain the industry-accepted maintenance work management structure and key elements. Your goal is to get to know your coworkers, their function, and the operation of their functional departments to create a process that benefits all stakeholders.

I always smile when I overhear a stakeholder say, "Let me show you how our process works" instead of "Let's look at Dale's process." This tells me that the team has taken ownership of the process and that the chances of sustainability are high.

CHAPTER ONE

PROJECT MANAGEMENT

MAINTENANCE WORK MANAGEMENT PROJECT STEERING COMMITTEE

A project steering committee is a high-level stakeholder group that provides strategic direction to the project and supports the project manager. It is important to emphasize that the group's priority is the success of the project.

Knowing that maintenance work management (MWM) impacts the frontline stakeholders and needs their buy-in to be successful, the successful project manager needs the support of upper management. This takes a project manager who is a knowledgeable MWM subject-matter expert (SME). I haven't encountered a formal MWM training program. With this book, you can understand the science and become an objective, credible MWM SME. It is the job of the MWM SME to educate and support the steering committee members while acting as the liaison for the process design team and frontline stakeholders.

Responsibilities of the Steering Committee

The steering committee approves the tasks and schedule for the project as defined by the project manager/MWM SME. The committee resolves conflicts between stakeholder groups, as the various groups help define and adhere to the completed, new MWM processes. Regular monitoring of the project's progress against the schedule is accomplished between the steering committee and the project manager at regular intervals.

The steering committee is responsible for the promotion of positive communication about the project and its benefits within the organization. The committee reviews and approves changes to the project schedule and their impact on safety, schedule, and cost and adjusts when necessary. Project pace will vary based on the organization's size and ability to adapt to change and improvement. Reasonable flexibility of all stakeholders is key.

Typical Steering Committee Members

Organizational structure varies widely from organization to organization for more reasons than can be listed. Functionally the steering committee is typically sponsored and led by the senior plant manager and includes an operations manager, a maintenance manager, and a project manager/MWM SME who also acts as the lead process designer, trainer, and coach. Throughout the MWM project, other functional experts may be added for advice as needed.

MAINTENANCE WORK MANAGEMENT PROCESS DESIGN TEAM

> *The MWM process design team is a team of frontline stakeholders who are responsible for defining the details of the six MWM processes—work identification, work approval, work planning, work scheduling, work execution, and work close out—and the responsibilities described in the MWM processes for the six processes led and coached by a knowledgeable MWM SME.*

When I was a very young manager, I was hired as a production control supervisor at a medium-sized, very well-known 100-year-old manufacturing company. After previously spending 8 years with a well-oiled global manufacturing operation that was well over 100 years old as a machine operator, I was shocked to find no job descriptions for the 18 employees on my team. In the first week, I met my team and asked each member to write me a list of the tasks they performed. In week 2, I reviewed their lists with them and asked them to write the steps to complete each task. In week 4, I presented to my middle-management colleagues "our" written departmental manual with job descriptions, processes, and a service-level agreement designed to lessen the conflicts and hard feelings, as my team was a service team.

As taught by my MWM mentors, it's "boots in the dirt." I've seen processes written by well-meaning managers and administrators only to fail during implementation or shortly after when the responsible and accountable frontline stakeholders didn't achieve buy-in because only they knew all the challenges that stopped them from being successful—challenges that could have been addressed during the discovery phase while identifying and analyzing the current as-is MWM processes.

Don't be afraid to use and enhance what you already have. During the discovery phase, search far and wide for "pockets of excellence" that already exist in your organization. We all know those among us who go far beyond the call of duty to be the best. Search every functional department to find what exists, and incorporate it into your process if relevant. After linking the pockets of excellence together, you will find your job to be a little easier just filling the gaps and tying it all together with an understandable process structure.

If you want buy-in, sustainability, and success, trust your team, and coach them by helping you to write your MWM processes to ensure your continual improvement and sustainability well into the future.

Process Design Team Objectives

Utilize design team members who are recognized as experts in their discipline, who see a need for change, and who are willing to act as change agents. Assess and identify gaps within the current MWM process. Produce required documentation: process flowcharts, step definitions, RACI charts, posters, training materials, and manuals of the MWM process. Provide the training and coaching necessary to ensure successful implementation. Provide ongoing coaching and auditing to ensure future sustainability.

Process Design Team Activities

It's very important that the design team members are knowledgeable, and most importantly, are devoted to positive change. Don't exclude team members because of their personalities. Look past personalities to the contributions made by members who are passionate about their work.

The members must be fully dedicated and participate in scheduled process design team meetings as SMEs in their discipline. They will act as liaisons in their areas of expertise to gather input from other stakeholders in the organization to identify best practices and existing tools for inclusion in the new MWM processes. They will assist with the collection and design of supporting processes and new tools for inclusion in the new processes. They become champions of the new processes within the organization.

The design team members will be expected to assist other team members during the rollout and training phase of the new MWM processes and provide future ongoing MWM coaching within the organization.

Typical Design Team Members

As discussed earlier, the project manager/MWM SME leads the team, which is typically made up of an operations second-line supervisor and an operations first-line supervisor, a maintenance second-line supervisor and a maintenance first-line supervisor, a maintenance planner (or two), and a maintenance scheduler.

As and when needed, also include an operations and maintenance coordinator for departmental tie-ins, a reliability engineer to define reliability data collection, and/or a CMMS SME or a supply chain expert. Never be afraid to include an expert when needed. This helps with acceptance of the new MWM processes and can produce other advocates.

MAINTENANCE WORK MANAGEMENT STANDARDIZATION

Dating back to the mid-nineteenth century, manufacturing management has recognized that standardization and repeatability of work increase productivity and decrease cost. MWM requires a commitment to fixing the systems and processes in a proactive way, not fixing the problems. Well-trained frontline stakeholders will fix the problems after taking part in the design, implementation, and utilization of optimized MWM processes.

When standardizing your processes, do the right things right. Strive to be effective, efficient, and proactive rather than reactive. Have a clear, purposeful focus with your philosophical maintenance approach. Ask, *Why are we doing this?* Eliminate activities where you cannot clearly define a purpose.

Have a maintenance vision and mission. Define what *good* looks like and how to accomplish it. Only consider tangible, measurable effects in:

- **SAFETY.** We all go home safe.
- **QUALITY.** We produce the best products on the market.
- **PRODUCTION.** We control our business and profits. We meet our planned goals.

MAINTENANCE WORK MANAGEMENT PROJECT PLAN

The project manager/MWM SME typically builds the project plan and is responsible for leading the MWM steering committee and MWM design team through the progress and reporting at regular intervals. Adjustments to the plan are negotiated with, and approved by, the MWM steering committee and MWM design team.

The typical major phases of the project plan include:

- 2 weeks for MWM steering committee and design team formation.
- 6–12 weeks for process design and the collection of existing supporting MWM elements. This includes work order/notification standards, risk

analysis tools, planning standards, and meeting agendas, every one to two weeks designing one of the six MWM processes, two to four weeks for process rollout, and 16–34 weeks for coaching and sustainability activities based on the size of the organization.

Figure 1.1 shows a typical MWM project plan for a small organization.

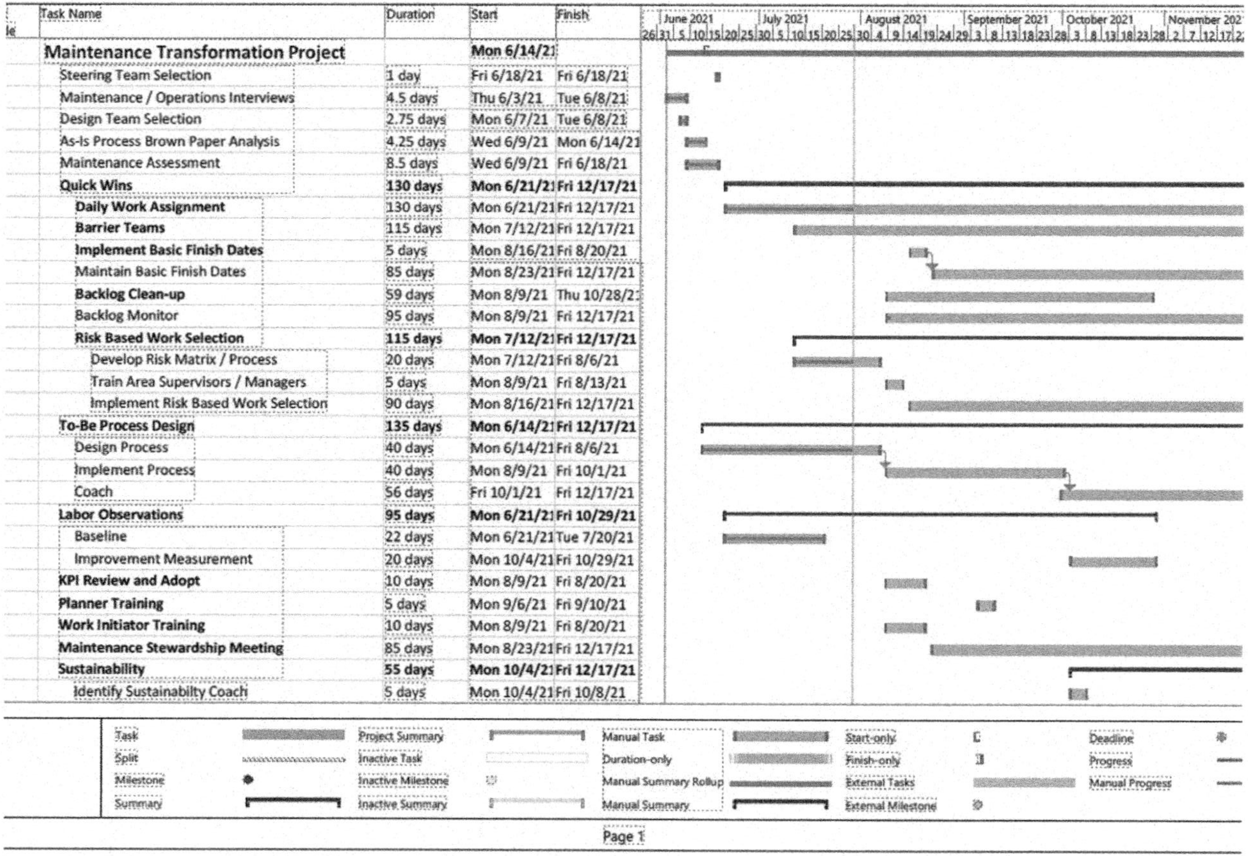

Task Name	Duration	Start	Finish
Maintenance Transformation Project		**Mon 6/14/21**	
Steering Team Selection	1 day	Fri 6/18/21	Fri 6/18/21
Maintenance / Operations Interviews	4.5 days	Thu 6/3/21	Tue 6/8/21
Design Team Selection	2.75 days	Mon 6/7/21	Tue 6/8/21
As-Is Process Brown Paper Analysis	4.25 days	Wed 6/9/21	Mon 6/14/21
Maintenance Assessment	8.5 days	Wed 6/9/21	Fri 6/18/21
Quick Wins	**130 days**	**Mon 6/21/21**	**Fri 12/17/21**
Daily Work Assignment	130 days	Mon 6/21/21	Fri 12/17/21
Barrier Teams	115 days	Mon 7/12/21	Fri 12/17/21
Implement Basic Finish Dates	5 days	Mon 8/16/21	Fri 8/20/21
Maintain Basic Finish Dates	85 days	Mon 8/23/21	Fri 12/17/21
Backlog Clean-up	59 days	Mon 8/9/21	Thu 10/28/21
Backlog Monitor	95 days	Mon 8/9/21	Fri 12/17/21
Risk Based Work Selection	**115 days**	**Mon 7/12/21**	**Fri 12/17/21**
Develop Risk Matrix / Process	20 days	Mon 7/12/21	Fri 8/6/21
Train Area Supervisors / Managers	5 days	Mon 8/9/21	Fri 8/13/21
Implement Risk Based Work Selection	90 days	Mon 8/16/21	Fri 12/17/21
To-Be Process Design	**135 days**	**Mon 6/14/21**	**Fri 12/17/21**
Design Process	40 days	Mon 6/14/21	Fri 8/6/21
Implement Process	40 days	Mon 8/9/21	Fri 10/1/21
Coach	56 days	Fri 10/1/21	Fri 12/17/21
Labor Observations	**95 days**	**Mon 6/21/21**	**Fri 10/29/21**
Baseline	22 days	Mon 6/21/21	Tue 7/20/21
Improvement Measurement	20 days	Mon 10/4/21	Fri 10/29/21
KPI Review and Adopt	10 days	Mon 8/9/21	Fri 8/20/21
Planner Training	5 days	Mon 9/6/21	Fri 9/10/21
Work Initiator Training	10 days	Mon 8/9/21	Fri 8/20/21
Maintenance Stewardship Meeting	85 days	Mon 8/23/21	Fri 12/17/21
Sustainability	**55 days**	**Mon 10/4/21**	**Fri 12/17/21**
Identify Sustainabilty Coach	5 days	Mon 10/4/21	Fri 10/8/21

FIGURE 1.1 A typical MWM project plan for a small organization. (Microsoft® Project Screenshot)

CHAPTER TWO

RELIABILITY AND MAINTENANCE

ASSET RELIABILITY

I've always known that reliability means not getting stuck at the side of the road while enjoying the full benefits of my equipment. This takes an ongoing investment that anticipates breakdowns by monitoring equipment health and fixing or mitigating anticipated failures. Reliability and preventive maintenance are an extensive subject and deserve a book of their own. In this book, we talk about the frontline interactions to ensure that the proactive work is done on time to prevent future failures and to collect the data necessary for analysis and improvement to achieve a high level of reliability.

Here are seven basic assumptions to keep in mind as you read on:

- Maintenance work management is where reliability meets hands-on asset care.
- Reliability and maintenance are not the same. Good maintenance, data collection, and analysis support asset reliability.
- Reliability is not a project; it's a continuous, measurable effort that only ends when the life of the asset ends.

- Reliability touches all stakeholders who use the maintenance work management process and the data it produces.
- Reliability protects production. Proactive reliability-centered maintenance anticipates and mitigates failures before they occur.
- Reliability is essential to decision-making. From production planning estimates to maintenance investment, asset uptime provides the financial and physical means to produce the organization's products.
- Reliability is an investment, not a cost. If you don't proactively maintain your assets, they will eventually fail.

Reliability-Based Maintenance Elements

A well-functioning integrated maintenance work management process, built on the six major processes, provides the solid foundation for all reliability and maintenance management. Figure 2.1 shows a managing system based on standardized maintenance work management.

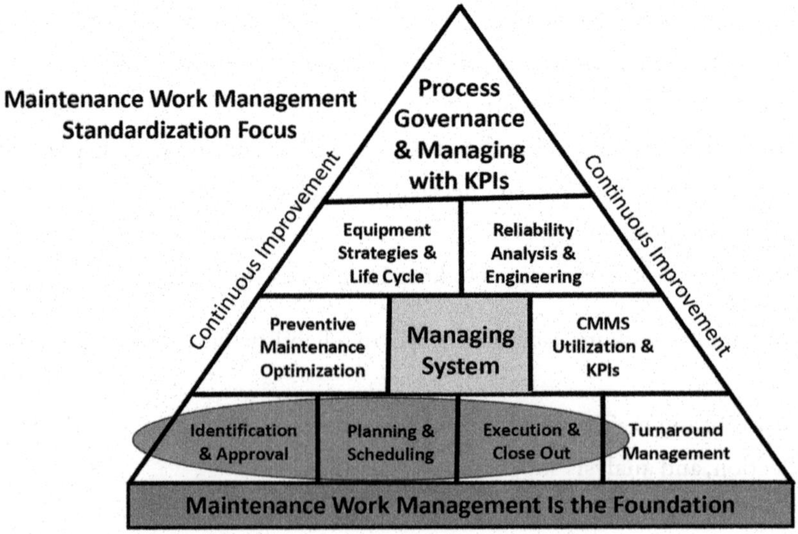

FIGURE 2.1 Maintenance work management standardization focus.

ASSET RELIABILITY INFLUENCES

Reliability takes into account plant configuration and equipment, maintenance process improvement, adjustment of operating procedures, training, spare part specifications and quantification, new plant design optimization, troubleshooting and diagnostics, and management guidance.

Production Considerations

Equipment reliability is supported by a closed-loop failure elimination system, a business reliability-driven preventive maintenance (PM)/predictive maintenance (PdM) program, reliability re-engineering, maintenance work management (MWM), planning and scheduling, materials optimization, and condition assessment and restoration.

A process reliability-based system consists of integrated process management, a process simplification and rationalization mindset, plant modernization, waste reduction, workflow reengineering, supply chain optimization, and process safety management.

Organizational effectiveness is achieved through integrated policy deployment, understandable organizational design, technical training solutions, plant industrial engineering, and well-structured change management support.

Business effectiveness can be optimized using a standardized opportunity assessment, benchmarking, automation, strategic outsourcing, computerized maintenance management systems (CMMSs), a contractor utilization and consolidation strategy, well-engineered manufacturing execution systems, and responsible processes ownership.

Unscheduled Downtime Cost

During unscheduled downtime, there is an escalation in labor costs and overtime as technicians sit idle between tasks until the resumption of production. They continue to be paid even as output and productivity come to a halt. Overtime may be necessary for complex repairs to make up for lost production.

Reactive repair costs are higher than planned maintenance costs due to the lack of planning caused by the urgency of the work, with more reliance on trial and error. Maintenance frontline supervisors and technicians have to identify and troubleshoot the failure, and then have to plan the work and obtain needed materials and components before repairs can be made. Other ongoing variable costs are incurred while running a production line, as overhead and other fixed expenses are still accruing.

Production Output Loss

Production rates can suffer losses through decreased run rates, as asset degradation leads to eventual failure. As a consequence, product quality may not adequately meet plant standards, resulting in disposal of inferior-quality output.

Excess Inventory

Buffers for on-hand tools and spare parts are inflated as protection against unexpected machinery failure and breakdown, with excessive carrying cost that is associated with the extra inventory. In addition, excessive schedule break-in work adds to an increase in expedited materials costs and shipping.

RELIABILITY AND MAINTENANCE WORK TYPES

Maintenance work is identified through two main categories: preventive work, which anticipates and mitigates future failure, and corrective work, to fix the equipment after it has started to fail. The preventive work encompasses any planned and repetitive work that will be completed on a set frequency utilizing preventive maintenance (time-based PM) or predictive maintenance (condition-based PdM) equipment strategies. Corrective work is as-found work needing repair. This is work that is not repetitive and may be due to equipment degradation or failure.

Corrective Work and Incipient Failure

> "An imperfection in the state or condition of hardware such that degraded or catastrophic failure can be expected to result if corrective action is not taken."
> —Center for Chemical Process Safety

It *will* fail once the failure has started! Enter a work notification *as soon as* you see the start of a failure!

Once a failure starts (P), monitor the degradation of the symptoms, and plot the results on a timeline, as shown in Figure 2.2. After a few measurements, plot a descending curve that reflects the measurement that indicates the functional failure point (F) in time when the asset no longer functions as designed. Set a timeline for the repair of the asset before the expected functional failure date.

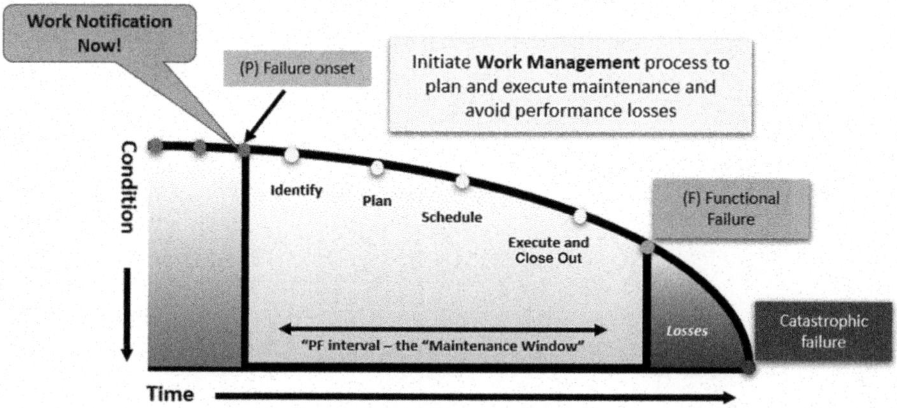

FIGURE 2.2 P-F curve/work order priority/target finish dates.

Preventive/Predictive Maintenance

Overall equipment effectiveness (OEE), which measures how well equipment/assets are utilized as compared with their design capabilities, is improved through asset availability, run rate performance, and product quality. Extended time between scheduled downtimes provides increased asset availability and increased craft productivity for scheduled work and minimizes costly asset damage.

Proactive asset management provides an integration with operations, reliability, and maintenance, safeguarding critical assets and satisfying regulatory compliance.

Maintenance by Operations

Also worth mentioning, maintenance by operations involves production personnel performing routine, minor maintenance tasks such as cleaning, inspection, lubrication, and simple adjustments to ensure equipment reliability. This proactive approach reduces downtime, prevents minor issues from becoming major failures, and improves OEE. Maintenance by operations is a method that I've seen employed during the early stages of work identification to designate minor work that does not need a notification, work order, reliability data collection, or maintenance resources, bypassing the MWM processes.

CHAPTER THREE

BECOMING AN EXPERT

MAINTENANCE WORK MANAGEMENT PROCESS DESIGN, IMPLEMENTATION, AND COACHING SUBJECT-MATTER EXPERT

The design, implementation, and coaching maintenance work management (MWM) subject-matter expert (SME) guides individuals and/or teams through the adoption of the MWM standardization process to help ensure individual and organizational MWM performance and success. The MWM SME team naturally progresses to the maintenance work management and sustainability governance body when the design and implementation project is complete, providing future expertise to the entire organization.

Engaging Key Stakeholders

> *Trust is the belief, expectation, and confidence that you are looking out for the other person's interest. Your goal is to increase motivation while working to understand and acknowledge the stakeholders' perception of risk.*

Trust is necessary to promote a solid relationship, which is necessary for implementation success; it must be established to outweigh real and perceived risk and be actively managed. Show that you are focused on the stakeholders' success; show that you are an MWM expert and that you are credible.

MWM Implementation Objective

Maintenance and operations organizations are very busy and are integral to keeping assets running to meet production goals. MWM is a process that will touch all aspects of operations and maintenance. It's very important that the MWM transformation does not interfere with day-to-day operations. The design and implementation phases must minimize disruption to the daily operations and activities, providing prescriptive guidance for key MWM process activities while also providing on-the-job coaching and support for all MWM activities.

The MWM project must maximize sustainability of process execution and performance, equip local resources to implement MWM changes across sites and regions, and provide tools to enable effective monitoring of process performance. As part of the project, define performance expectations and goals, and manage the MWM process execution to the established standards. Assess and evaluate implementation issues quickly and decisively, instituting changes and/or corrections to address concerns as they arise.

Identify and Leverage Opportunities

Listen to the stakeholders at all levels to demonstrate and model the desired behaviors; coach to performance in a nonthreatening way. Remove barriers to success by improving process understanding and usage. I always say, "Sell it; don't tell it." Escalate questions and issues as necessary. As an MWM SME, never stop learning.

MWM SME Activities

It will sometimes seem that the job of an MWM SME is boundless. An MWM SME's activities include making on-the-job observations, looking for coaching opportunities, conducting work reviews, examining MWM adherence on a regular basis, and attending regular work-related meetings to discuss issues and concerns with process design, tools, and performance.

MWM SMEs provide detailed instructions as required to reinforce training concepts, and they report progress on the individual stakeholders at the functional and process levels. They develop improvement plans and rewards based on the individuals' progress and structured activities to develop appropriate capabilities and recognition.

MWM SME Key Skills

The skills employed by the MWM SME can seem endless. Don't be intimidated—just be yourself and stick to the MWM science. MWM SMEs establish trust by being empathetic and understanding the stakeholders they engage with. They question and listen, letting the stakeholders be part of the process while they consult, giving advice only when needed after clarifying issues and understanding the stakeholders' concerns.

MWM SMEs make decisions focusing on the process, not personal opinions, and anticipate future problems identifying resistance to process acceptance. They manage team involvement by encouraging every stakeholder to take part in the problem-solving process, listening to all inputs and drawing logical conclusions.

MWM SMEs manage the project, monitoring and adjusting as needed to a written project plan. They manage meetings, establishing and adhering to meeting agendas. They manage conflict and varying personal styles by remaining neutral and also returning to the process for conflict resolution through negotiation, remaining objective and exhibiting a willingness to compromise while staying aligned with the processes.

MWM SMEs manage human performance, evaluating stakeholder performance to the processes and correcting deviations by selling the processes, knowing that they can convince stakeholders to change, but not force them to change. They diagnose the need for intervention and develop required corrective actions using their knowledge of the MWM processes, then coach by identifying the need for change and using the processes as the basis for change.

By using their industry knowledge, technical skills, and understanding of the operation of the physical assets, the MWM SMEs design solutions within the MWM processes using inputs from all stakeholders to develop one integrated process, fit for all stakeholders.

MWM SMEs support the use of the processes and help individuals and groups to resolve issues and remove the barriers for success, providing knowledge and expertise while spreading process knowledge from the sideline, not in the day-to-day performance of individual roles. They do not take ownership of individual and group issues, act as a clerk to type up meeting notes, work as a "pair of hands," solve everyone's problems, or do others' work for them. The SMEs' daily role after implementation is to observe performance to the processes and coach the stakeholders to perform to the processes. This includes stakeholders at all levels of the organization.

MWM SME Interpersonal Skills

MWM SMEs have the ability to develop relationships quickly and effectively, asking open-ended questions to quickly collect relevant information. They actively practice their listening skills for building trust. They ensure the quality of information. MWM SMEs have the ability to work within a range of styles and personalities by managing conflict.

MWM SME Coaching

Coaching is what we say and do informally to help people with their performance on a day-to-day basis to provide an environment that allows stakeholders to monitor and continually improve their performance.

As a coach, we recognize legitimate opportunities to coach, set the stage for coaching, agree on the setting, develop multiple solutions, agree on a plan of action, and follow up to recognize accomplishments.

Effective Coaching

Always focus on process performance, not the stakeholder. Aim for improvement and the pursuit of excellence, not perfection. Consider the stakeholder's preferred communication style when tailoring your message. Involve the stakeholder when seeking a change or developing a new idea.

Use a collaborative, not a controlling, approach. Accentuate the positive, and focus on the path around obstacles. Build trust as the foundation of a good coaching relationship. Focus on consistency, caring, commitment, and capability. Capitalize on coachable moments as they arise during normal day-to-day activities.

MWM Process Coaching Opportunities

Coaches engage throughout the entire MWM design, implementation, and sustainability phases of the project by working with stakeholders impacted by the work identification, work approval, work planning, work scheduling, work execution, and work close out processes. They work with supervisors and managers to improve process performance and address issues.

Coaches attend process-specific and work performance meetings, engaging with participants and interacting with meeting leaders to execute planned agendas and mediate or resolve issues. They work with process owners to monitor process effectiveness and performance, addressing and resolving process issues. This includes process interfaces engaging with other related business processes, e.g., accounts payable, supply chain, reliability, engineering, and projects.

Tips for Coaching

Start at the earliest point in the process wherever improvement is important. Be subtle. You know what will be coming later in the project and application of the maintenance work management science. Without giving away what's to come, guide others toward best practices that already exist. Leverage the best practice owners' knowledge to help you gain acceptance of change.

Always be looking to identify the coaching point, and discuss actions to maintain a good process and make corrections to process flaws. Constantly test for understanding and agreement by using active listening techniques. Focus on major process issues. Don't get caught up in minor issues that won't impact process performance. Guide stakeholders to refine their work based on constructive, process-based feedback.

MAINTENANCE WORK MANAGEMENT PROCESS DESIGN, IMPLEMENTATION, AND MANAGEMENT

As noted earlier, the six major processes of MWM are work identification, work approval, work planning, work scheduling, work execution, and work close out, as shown in Figure 3.1. Each process is defined in the list below, along with other MWM definitions:

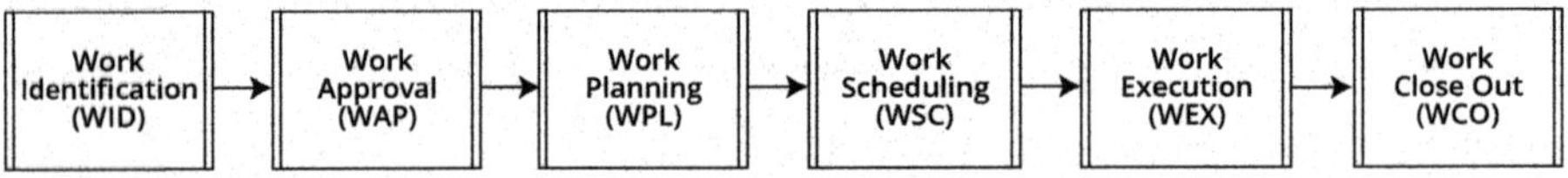

FIGURE 3.1 Maintenance work management standard six processes.

Maintenance Work Management Definitions

- **WORK IDENTIFICATION.** Recognizing that maintenance resources are needed to perform work to safely maintain assets/equipment and protect production and quality, while documenting that need in a notification and/or work order (CMMS) to the quality notification standard.
- **WORK APPROVAL.** Verifying the quality of all notifications or work orders, setting the priority and required end date or target finish date, ensuring all work will be performed when needed, before approving the work to progress.
- **WORK PLANNING.** Determining and documenting the tasks and activities needed to safely accomplish work, including estimating

labor resources, materials, tools, equipment, information, and safety requirements.

- **WORK SCHEDULING.** The synchronization of all resources needed at a particular time and place to accomplish the work across multiple work orders simultaneously.
- **WORK EXECUTION.** Completing the work safely per the work plan and documenting the information identified in the quality work order standard.
- **WORK CLOSE OUT.** Completing the planned work and documenting the information identified in the quality work order standard, creating relevant histories (CMMS).
- **KEY PERFORMANCE INDICATORS (KPIs).** Reporting values created for given objectives such as schedule compliance, schedule break-in work, planned work, planned versus actual cost, and budget variances.
- **ANALYSIS AND IMPROVEMENT.** Reviewing the work completed, ensuring that effectiveness of the processes and execution performance are analyzed for lessons learned and that any changes needed to improve work plans and processes are implemented.
- **HISTORY.** Capturing and recording relevant historical data for the equipment/asset and work executed, after the work order has been completed, to be used for analysis and improving reliability.

CHAPTER FOUR

ANALYZING THE CURRENT MAINTENANCE WORK MANAGEMENT PROCESS

BROWN PAPER ANALYSIS

A brown paper analysis is a very effective way to build a representation of the perceived existing processes and is best conducted by a cross-functional team of the frontline stakeholders. We want to identify the current processes, including work identification, work approval, work planning, work scheduling, work execution, and work close out, while identifying the stakeholders' perception of the processes gaps, failures, responsibilities, and accountabilities.

This team is best served using the design team and expanding it to include others holding the same functional positions and adding representatives from the reliability, supply chain, and CMMS teams if available. We want all these people in the room at the same time to spark a constructive, in-depth conversation and analysis. Remember, one major goal of maintenance is collecting data that will increase and maximize asset reliability and process integrity.

Building the As-Is Process

In a process flowchart format, start with the major processes—work identification, work approval, work planning, work scheduling, work execution, and work close out—to structure your analysis as titles across the top of the flowchart. For each of the major processes, build the next-level detailed flowcharts, identifying the individual process steps and relationships between them using sticky notes. Remember, this is your current process as the participants perceive it. Resist making process improvements at this stage.

Areas and Opportunities to Engage In

Consider all process and content gaps. Ask, *Does our current perceived process address all the elements of industry standard MWM?* Think about organizational gaps. *Do we have positions or functions in our organization that match the six MWM processes?* Identify subprocess gaps. *Do we have process details and tools to satisfy all steps in our perceived MWM processes?* Consider skills gaps. *Do we have the needed skills to design and utilize a process that fills our gaps?*

The next step is to document the gaps and areas for improvement. In the following example, red clouds show gaps and areas for improvement to a dramatic effect. Use these red cloud statements to build a list of gaps and areas for improvements to be used when designing your new MWM process. By eliminating all the red clouds, you will satisfy the concerns of all stakeholders during the design phase of your new MWM processes. You can find many process brown paper analysis examples and instructions online, and you can see what one looks like in Figure 4.1.

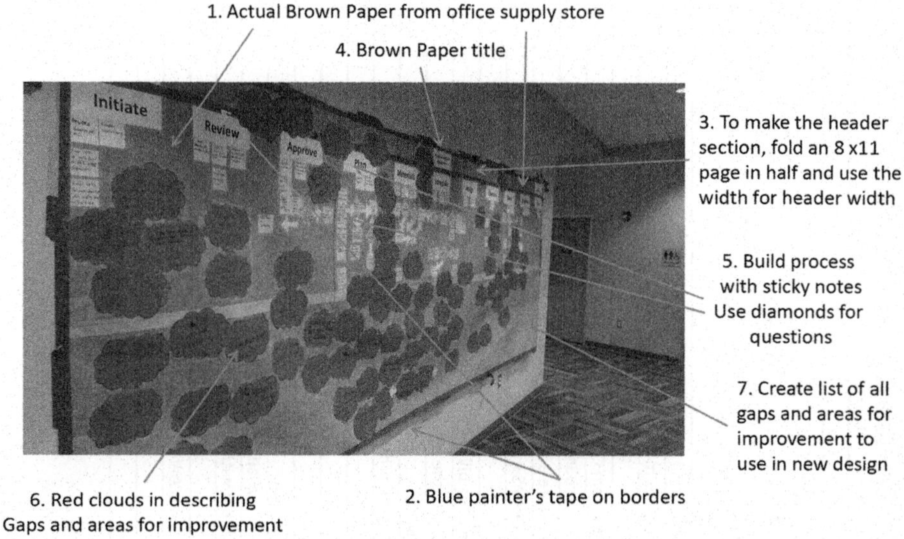

FIGURE 4.1 As-is process of Brown Paper analysis.

Red Cloud List

Use the red cloud statements to build a list of gaps and areas for improvements to be eliminated when designing your new MWM process, identifying who's accountable (functional area), process activity (six process flowchart titles), and the verbiage from the red clouds. Figure 4.2 provides an example of a Red Cloud List.

Accountable	Process Activity	Red Cloud
Operations	Work Identification	When needing maintenance help, call supervisor, not putting in CMMS
End-to-End Process	Work Identification	Pipeline hierarchy not well defined
Operations	Work Identification	I&E PM work tracked on spreadsheet, not in CMMS
Operations	Work Approval	Work order review needs to be more detailed
End-to-End Process	Work Approval	Lack of backlog visibility/management
Maintenance	Work Approval	After approval should go to scheduler/planner for plan and materials, then to supervisor or craft lead
End-to-End Process	Work Approval	Backlog management in risk analysis needed
Maintenance	Work Execution	Don't know if the work order is in CMMS when it is assigned
End-to-End Process	Work Execution	Need to follow work processes and use CMMS more
Operations	Work Execution	Field operators are not trained in CMMS
Maintenance	Work Scheduling	Outlook base schedule built through Excel from CMMS
Maintenance	Work Planning	Planner could use applied planning and scheduling skills training
End-to-End Process	Work Planning	Needed to perfect the added processes before adding more
End-to-End Process	Work Planning	Need follow-up work processes and use CMMS more
Maintenance	Work Planning	No library of standard jobs
End-to-End Process	Work Planning	Need to learn new skills for roles and responsibilities
Maintenance	Work Planning	Get new planner/scheduler up to speed with the process
Operations	Work Execution	Field operators doing work, not using CMMS
Maintenance	Supply Chain	Planner not always putting materials on work orders
Operations	Supply Chain	Field operators buying parts
Purchasing	Supply Chain	Centralized warehouse coming soon
End-to-End Process	CMMS Data	Online
Maintenance	CMMS Data	Trades complaining about too much paperwork
Reliability	CMMS Data	Data lacking but starting to use what data is available
End-to-End Process	CMMS Data	PM workflow is developed and implemented without feedback from affected group
Maintenance	Work Execution	Add labor on complete step
Maintenance	Work Execution	Supposed to review actuals and failure codes before completing work orders
Reliability	Work Execution	Need effective review process
CMMS	Work Close Out	Not sure who closes work order

FIGURE 4.2 An example of a Red Cloud List.

"DAY IN THE LIFE OF" STUDY

The Day in the Life Of (DILO) study is the measurement of a key administrative, supervisory, or management stakeholders' hourly activities that support the use of the maintenance work management (MWM) processes. In many cases, it's the analysts first introduction to these key stakeholders and their current perceived responsibilities. It's a powerful learning that reaches into the detailed operations of the organization.

It's a great opportunity for the analyst to get to develop a trusting relationship with the key stakeholders who will be involved with the design, implementation, and sustainability of the new MWM processes. It identifies personal concerns for further study, consideration, and elimination when developing the new processes.

The studies should be scheduled at times that do not interfere with the stakeholders' performance of their normal duties. It's helpful to observe as much of the stakeholders' typical day as possible. With growing trust, it's possible to coach the stakeholders to outline their activities that the analyst is unable to observe. After several DILOs with various stakeholders, the analyst can begin to understand the interactions between the stakeholders and the current as-is MWM processes.

The operations frontline supervisors are also key to the MWM process. On the front end of the MWM processes, they are typically responsible for work identification and approval as the owner of the operating and maintenance budgets. In the middle of the MWM processes, they are instrumental in the maintenance scheduling and daily work assignment processes with the authority to approve and break the maintenance schedule. Be aware that their role extends far beyond their maintenance responsibilities. An MWM DILO on an operations frontline supervisor may only involve a small percentage of the supervisor's total workday.

When observing the operations frontline supervisor, be aware of the supervisor's responsibilities and activities to support the MWM processes. Ask, *Does the supervisor approve all new notifications daily? Does the supervisor attend the daily work assignment meetings? Does the supervisor approve the weekly maintenance schedule? Does the supervisor decide what work will be deferred to accommodate work that breaks into the maintenance schedule?* Incorporate the supervisor's non-MWM activities in the study.

The DILO Process

Select the key stakeholders for the study. Be familiar with the MWM processes and the stakeholders' responsibilities in the processes. Make sure to select a broad enough number of studies to get observation across the entire workday to ensure that you are not limiting yourself to "busy times" or "quiet times."

Try to establish industry standards for the percentage of the overall time they use for the total of their individual tasks; i.e., the maintenance planner plans work 75% of their time, the maintenance frontline supervisor is in the field observing the execution of work 60% of their day, etc. See Figure 4.3.

Ideal Planner Time Distribution	
Field Scoping / Research	30.0%
Plan, Sequence, Material (CMMS)	35.0%
Maintenance, Production, Engineering Interface	10.0%
Administration / Meetings	15.0%
Non-Planner Work	0.0%
Away or Non-Value Added	10.0%
Total Planning Activity	75.0%

FIGURE 4.3 Maintenance planner ideal time distribution.

Maintenance planners are typically very knowledgeable key stakeholders who are depended on too heavily by other stakeholders due to their vast body of knowledge and understanding of the maintenance and reliability functions. The complexity of the planning process is not well known, and the maintenance planners are too often viewed as administrators, responsible for expediting all the requirements of job execution at the correct time. This is combining maintenance planning, scheduling, and materials into a short interval of time.

When observing a maintenance planner, understand the importance of the maintenance function. The combined planning group must plan enough hours to utilize all technicians 100% of the time. This requires a large number of detailed plans and typically takes 75% of the planner's time to field-scope the work; build the tasks to complete the work; identify the needed labor, information, materials, and components to complete the work; and once the plan is approved, set these activities in motion to hand off to the maintenance scheduler.

Many organizations physically separate the maintenance planners from the maintenance frontline supervisors, with a total separation from the day-to-day execution of maintenance work, so they can maintain enough planned work to consume the maintenance technicians for six to seven weeks.

When observing a maintenance frontline supervisor, ask, *Is the supervisor spending 60% of the day in the field observing the execution of work by the technicians? How much time is the supervisor spending preparing and assigning work to keep the technicians working to optimize labor cost? How much time does the supervisor spend in required meetings and administrative duties? How much time does the supervisor spend performing the duties of other key stakeholders?* See Figure 4.4.

Ideal Maintenance Supervisor Time Distribution	
Training	5.0%
Direct Supervision	25.0%
Indirect Supervision	35.0%
Administrative Duty	25.0%
Hourly Work	0.0%
Away or Non-Value Added	10.0%
Total Supervisory Activity	65.0%

FIGURE 4.4 Maintenance frontline supervisor ideal time distribution.

DILO Worksheet Templates

Develop DILO worksheet templates for the various key stakeholder roles with responsibilities from the MWM processes. Keep in mind that direct labor technicians completing the maintenance tasks are measured using the time-on-tools (ToT) process. As you conduct your observations, add new tasks to your form that may not meet the responsibilities of the MWM processes. Figures 4.5 and 4.6 provide examples of maintenance planner and front line supervisor data entry templates.

You will see in this book that complicated software programs are helpful but not necessary. All the tools demonstrated can be built in popular desktop applications with basic knowledge. To avoid carrying and using a laptop during the study, I take notes in my journal and enter my observations later in my computer.

Routine Maintenance Planner Time Management

PLANNER	John Doe	DATE:	11-Feb-26
DEPARTMENT:	Mechanical Planner	OBSERVATION:	12:30 TO: 16:30
SHIFT:	Day	1st BREAK	TO:
GROUP SIZE:	1	LUNCH:	TO:
		2nd BREAK:	TO:

START	STOP	FIELD SCOPE/ RESEARCH	SEQ/MAT'L	MX/OPX/ ENG I/FACE	ADMIN/ MTNGS	NON-PLAN	AWAY/NVA	COMMENTS
12:30 PM	1:10 PM	30		10				Alky Gear; sign-on to unit; review WO's in unit. Review Crude
1:10 PM	1:30 PM	20						Field scope WO's
1:30 PM	2:00 PM	25					5	Travel; break; field scope WO's.
2:00 PM	2:05 AM						5	Personal Break
2:05 AM	2:20 PM		5	10				Vendor call; scheduler discussion; WO in CMMS
2:20 PM	2:30 PM	7		3				Weld Specs; Weld Quality Control Form
2:30 PM	2:50 PM		10	5		5		Seek L3 input on request
2:50 PM	2:55 AM					5		Phone call from vendor re: invoice issue
2:55 PM	3:15 PM		15			5		WQCF; Standard Job; build packages; scheduler discussion re: why job not complete - non planner discussion
3:15 PM	3:25 PM		2		5		3	Personal call
3:25 PM	3:35 PM		5	5				MIQA; BOM package; seek engineer for input
3:35 PM	3:45 PM			5		5		Vendor re: steam trap specs; engineer re: spec issue for tgrap
3:45 PM	3:55 PM		5		5			steam trap paperwork; e-mail
3:55 PM	4:05 PM		2		5	3		paperwork shuffle
4:05 PM	4:15 PM		5				5	finish CMMS; casual discussion with visitors
TOTALS	225	82	49	38	15	23	18	TOTAL PLANNING ACTIVITY 75%
% TOTALS	100%	36%	22%	17%	7%	10%	8%	

FIGURE 4.5 Maintenance planner data entry template.

Maintenance Frontline Supervisory Time Management

SUPERVISOR:	Jane Doe		DATE:	March 1, 2026	
DEPARTMENT:	Field Mechanic Supervisor - Bldg 2		OBSERVATION:		TO:
SHIFT:			1st BREAK		TO:
GROUP SIZE:	1		LUNCH:		TO:
			2nd BREAK:		TO:

START	STOP	TRAIN	DIR. SUPV.	ADMIN DUTY	AWAY/NVA	COMMENTS
6:00 AM	6:30 AM		30			Tool box
6:30 AM	7:00 AM			30		Contractor spill; ENV paperwork; caustic spill
7:00 AM	8:00 AM			60		Contractor spill; ENV paperwork; caustic spill
8:00 AM	8:30 AM	30				
8:30 AM	9:30 AM			60		Spill paperwork
9:30 AM	10:00 AM		30			
10:00 AM	10:35 AM		35			
10:35 AM	10:40 AM				4	
10:40 AM	11:00 AM		20			
TOTALS	299	30	115	150	4	TOTAL SUPERVISORY ACTIVITY 48%
% TOTALS	100%	10%	38%	50%	1%	

FIGURE 4.6 Maintenance frontline supervisor data entry template.

DILO Study Activities

Introduce yourself to the stakeholder, and describe the purpose of the study and how long it will take to complete. Make it clear that the purpose of the study is not to assess individual behavior but to assess how much of the stakeholder's time is spent on the activities from the MWM processes. Explain that personal time (e.g., lunch, bio breaks) will not be observed or assessed. Assure the stakeholder that the results are anonymous and confidential.

Ask the individual stakeholders to describe their role as they perceive it. This is crucial to being able to categorize activities. Take notes on their key responsibilities as they perceive them. Let the stakeholders know that you will try not to interrupt them or get in the way but that you might be asking a few questions along the way for clarification.

Record the actual start-to-finish times for each different task that a stakeholder performs. At the end of the worksheet, consolidate the total results for each category. Using the total duration for the entire observation, calculate the percentages spent on each individual task category.

As you are observing each task, take detailed notes to support the task, especially tasks that are not the stakeholder's responsibility per the MWM processes' RACI chart. These deviations will be vital in building your business case for change. Feel free to openly discuss with the stakeholder the different tasks that you observe while they occur and ask for justification for performing the tasks in a nonthreatening way. Always remember, relationships and trust are a valuable asset.

Ideal Time Distribution Versus Observations

After conducting and consolidating the observations, present the results in an easy-to-understand format. See Figures 4.7–4.10 for some examples.

Ideal Planner Time Distribution	
Field Scoping / Research	30.0%
Plan, Sequence, Material (CMMS)	35.0%
Maintenance, Production, Engineering Interface	10.0%
Administration / Meetings	15.0%
Non-Planner Work	0.0%
Away or Non-Value Added	10.0%
Total Planning Activity	75.0%

Planning (R/M and Outage)

Planner Time Management Composite (5) (R/M and Outage)	
Field Scoping / Research	15.0%
Plan, Sequence, Material (CMMS)	44.0%
Maintenance, Production, Engineering Interface	8.0%
Administration / Meetings	15.0%
Non-Planner Work	1.0%
Away or Non-Value Added	17.0%
Total Planning Activity	66.0%

Category	Description
Field Scoping/Research	Physical investigation of job site and requirements; review of historical data for job planning.
Plan, Sequence, Materials	Developing job estimates, standard jobs, operations, material needs, requisitions, etc.
Interface	Soliciting work scope/requirement input with operations, engineering, maintenance personnel.
Administrative/Meetings	Meetings, reporting, non-planning paperwork, personnel issues or activities, etc.
Non-Planner Work	Performing work typically assigned to supervisors, managers, wage employees, etc.
Away/Non-Value Added	Breaks, personal business or absence from defined responsibilities.

FIGURE 4.7 Maintenance planner time management ideal versus observations.

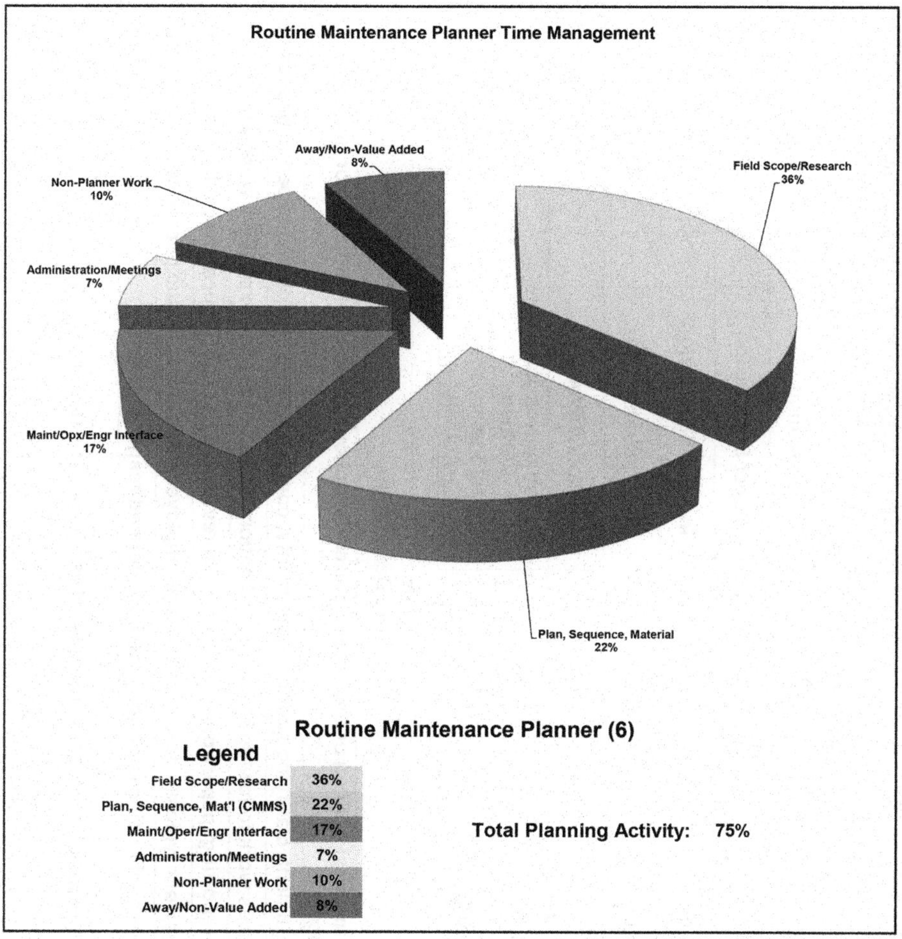

FIGURE 4.8 Maintenance planner time management composite pie chart.

Maintenance Supervision (R/M Only)

Ideal Maintenance Supervisor Time Distribution	
Training	5.0%
Direct Supervision	25.0%
Indirect Supervision	35.0%
Administrative Duty	25.0%
Hourly Work	0.0%
Away or Non-Value Added	10.0%
Total Supervisory Activity	65.0%

Supervisory Time Management Composite - 6 Supervisors R/M Only	
Training	6.0%
Direct Supervision	22.0%
Indirect Supervision	23.0%
Administrative Duty	43.0%
Hourly Work	0.0%
Away or Non-Value Added	6.0%
Total Supervisory Activity	51.0%

Category	Description
Training	Formal or informal training or one-to-one corrective action related to productivity or quality performance.
Direct Supervision	Work assignment, clarifying scope requirements, communicating completion expectations, etc.
Indirect Supervision	Encompasses work follow-up status, coordination of work activities, expediting work, etc.
Administrative Duty	Meetings, reporting, paperwork, payroll, formal reviews, personnel issues or activities, etc.
Non-Supervisor Work	Performing work typically assigned to wage employees or other staff/line personnel.
Away/Non-Value Added	Breaks, personal business or absence from defined responsibilities.

FIGURE 4.9 Maintenance frontline supervisor time management ideal versus observation.

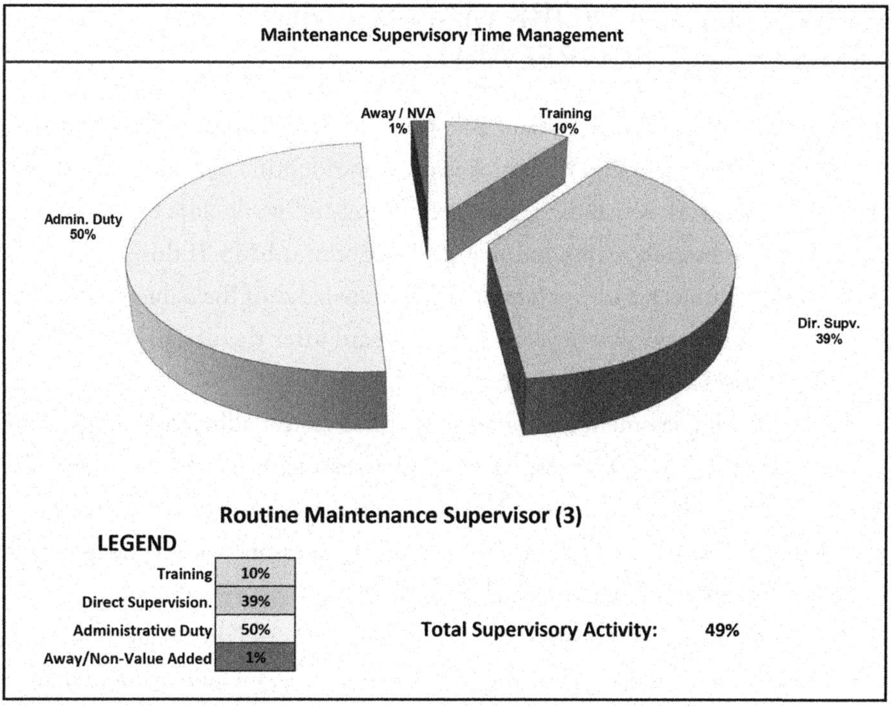

FIGURE 4.10 Maintenance frontline supervisor time management composite pie chart.

While doing your overall DILO study, if time permits, conduct observations with as many stakeholders within a particular job title as time permits to get a good cross section of the average MWM as-is processes' understanding and the stakeholders' adherence to the as-is processes.

After building the brown paper analysis and conducting DILO studies, you are beginning to develop an understanding of the current as-is MWM processes, the stakeholders' understanding of the processes, their adherence to their as-is MWM processes, and many valuable gaps that will help when designing the new MWM processes.

MAINTENANCE WORK MANAGEMENT ASSESSMENT SCORECARD

The maintenance work management assessment scorecard is tailored to the details of your organization's six MWM processes: work identification, work approval, work planning, work scheduling, work execution, and work close out, as well as the barrier elimination teams and maintenance stewardship. If this is your first MWM design project, and you have no prior knowledge of the industry standard MWM processes, you will develop this scorecard after the design of your new MWM processes and use it in future audits.

The MWM assessment sections are designed to test adherence and performance to the MWM processes. As with all assessments conducted internally, choose assessors who know and understand the processes and can objectively and fairly assign scores and suggest improvements. Use assessors who are independent of the MWM processes where available.

Create a numerical scoring system that gives extra weight to questions pertaining to the critical aspects of the MWM processes, including maintenance work risk assessment, work planning, work close out, and any other functions identified as critical to your particular organization.

For each assessment segment, check for a written process, posted process flowcharts, clear roles and responsibilities, process training delivered, and understanding of the process. Score and provide comments for all questions on the assessment. As part of the overall MWM assessment report-out, these comments are used to explain the score.

Work Identification and Approval

Test adherence to this process by selecting a sample of newly entered notifications. Check the quality of these notifications to a written quality standard.

Attend work screening meetings to test adherence to a documented written agenda if it exists. Look for adherence to established break-in criteria even if they are perceived. Pay particular attention to the use of a risk assessment tool for assigning work priority and required end dates.

Work notification content is very important for saving time for the MWM processes' downstream stakeholders. It's important to avoid wasted time when one of these stakeholders has to return to the work notification initiator to get more information or to clarify information in the work notification.

An example of a work identification and approval scorecard is shown in Figure 4.11.

Maintenance Work Management Assessment

Friday, June 25, 2021

Criteria	Pts	Score	Comments
Work Identification and Approval			
Is the work identification process documented? Is there clear understanding of the workflow? a. Is there a posted workflow chart? b. Have all supervisors been trained to the workflow roles & responsibilities?	10	5	A basic process flowchart exists indicating the roles and responsible for each step in the process. Interviewees indicate that they are not aware of the flowchart and that they have not been trained to the roles on the flowchart. No written procedures were observed.
Work Screening – Have the "Perfect Work Order" criteria been communicated to the business team Process Operators (or other work order quality criteria)?	10	5	Instructions exist for the use of SAP and the minimum of information required on a notification. Training is OJT. No written criteria for the content describing the information needed to plan the work.
Are work notification descriptions complete? -- Review a sample of 20 work orders that have been planned within the last 60 days. -- Review a current sample of work notifications.	10	2.5	The degree of the definition of the issues encountered when completing the long description in the Notifications varies greatly from defined to nonexistent. None of the samples were "well-defined." The short descriptions do not conform to established instruction for the use of SAP. There is no written quality standard or training to a standard.
Are Work Notifications being screened daily? Quality & Completeness.	10	5	Screened daily by Operations. Quality or content are by personal preference.
Work Screening – Meeting: a. Are work screening meetings being held daily? b. Is there a standing agenda, and is it being followed? c. Is the meeting duration 15 minutes or less? d. Did the meeting start on time?	10	2.5	Screening is completed by Operations Area Supervisors and Managers. No agenda or meeting exists. There is little to no structure and tracked input from Maintenance.
Are Hurdle Rates & Break-in criteria documented? Have Hurdle Rates & Break-in criteria been communicated to the business team?	10	0	No Risk Matrix, Hurdle Rates, or Risk-Based Work Selection criteria exist for setting Basic Finish Dates on Work Notifications. Basic Finish Dates are not being used to indicate future failure expectations or to trigger future risk analysis if work isn't completed on time..
Does break-in work exceed 5% of work orders completed? If more than 5%, what is the approx. % (> 10% = 5 pts, > 20% = 0 pts)	10	0	Interviewees report that break-in rate is between 25-50%. Weekly metrics for week 23, 2021 indicates 32% break-in.
Does the Business Team have a Break-in Stewardship meeting which includes Mechanical & Process representation (or are break-in reviews done to maintain compliance)?	10	0	Metrics are produced weekly and reviewed by individual team members. No current break-in review meeting organized and scheduled.
Have the work screeners completed Risk Assessment training?	10	0	Risk assessment is assigned per personal perception of risk with no Risk Matrix used to set Basic Completion Dates.
Are standard Service Deliverable Dates (SDD (standard lead-time to repair by failure and equipment class)) being used as the basis for selecting Basic Completion Dates for all work orders? Review sample of 20 work orders.	10	0	No SDDs exist by repair or equipment class. No Basic Finish dates are being used to set priorities for planned and scheduled work.
	100	20	

FIGURE 4.11 Work identification and approval scorecard.

Work Planning

Planning is vital to all business functions; MWM is no exception. Materials management is closely tied to planning. Pull a sample of planned work orders, and test the quality of the plan to established industry standards for breaking the work into tasks that identify the work to be completed, the labor resource needed, the number of labor resources needed, the duration of the task, the work instructions needed, drawings, special tools, equipment, technical information, and the required parts and materials.

Spend time with maintenance planners asking them to explain their processes. Ask, *How do you prioritize work? Do you look at the quality of the work notifications and return them to the work initiator if quality is poor? Do you visit the worksites to develop your plans? What are the elements of your plans? Do you get proper approval for your plans?*

An example of a work planning scorecard is shown in Figure 4.12.

Maintenance Work Management Assessment

Friday, June 25, 2021

Criteria	Pts	Score	Comments
Work Planning			
Is a Material Coordinator in place and functioning full-time?	10	5	Material coordination is handled by the planner, maintenance supervisors, technicians, and stockroom attendants. A full-time coordinator may not be necessary; however, the function can be defined and assigned to the appropriate department.
Are jobs being "stepped out" (detail tasks/subtasks) with estimated times (or unit rate) for both company and Contractor crafts? Review sample of 20 work orders.	10	5	PdMs and PVMs have the basics required to schedule, perform, and document the tasks, including tasks with resources identified and time durations to complete the tasks. The samples indicated that Corrective Maintenance Notifications rarely include tasks, resources, and time durations. The work orders reviewed indicate the minimum needed to charge labor time or compensate contractors. The maintenance supervisors are adding tasks when not planned.
Is the Planner removed from the day-to-day execution of work - Emergency jobs and expediting material?	10	0	By definition, no "planners" are in place. The existing Planner functions as planner, scheduler, contractor supervisor, coordinator, and expeditor. Planner is involved in the daily execution and the troubleshooting of work execution. The Planner fits the established description of a Maintenance Coordinator.
Does the Planner attend a weekly planning meeting (multi-craft jobs planning) each week?	10	5	Multi-craft planning is conducted by the planner and maintenance supervisors on an ad hoc basis. No weekly planning meeting is held to facilitate multi-craft planning. If needed, planner meets the contractor supervisor to schedule next week.
Are Planners field-scoping jobs and utilizing a compliant planning checklist?	10	5	Planner is field-scoping jobs when needed and is not utilizing a planning checklist. Planning checklist was not observed.
Are parts and material identified on the job plan? Sample 20 work orders within the last 60 days.	10	0	Parts and materials for Corrective work are identified via an attached BOM, not using the SAP Components tab. Minimal PVSs and PdMs require materials.
Are Simple, Basic, and Complex job plan criteria documented and utilized to determine needed documentation and procedures required for execution? Is the 3-month "look-ahead" schedule being used to prioritize the planning process for long-lead items?	10	0	Jobs are not classified as Simple, Basic, or Complex, and no criteria exist as such. No 3-month look-ahead schedule is being utilized in routine maintenance.
Is there stewardship in place to track planners' effectiveness (or planners' scorecard utilized)?	10	0	Planner effectiveness is not measured utilizing a standard scorecard or established KPIs.
Does the Maintenance organization or Business Team have a feedback process on the effectiveness of job plans produced - accuracy of estimates, all required steps, appropriateness of parts and materials identified? Is there feedback to planning on plan quality (plan vs. actual)?	10	0	Plan quality and planner feedback are ad hoc and not recorded or communicated in a formal process.
Are job packs being developed and properly utilized? Sample 10 job packs for content/quality	10	5	No formal job pack requirements exist. Job packs are minimal and provide only the basics needed to complete the job.
	100	25	

FIGURE 4.12 Work planning scorecard.

Work Scheduling

Work scheduling is vital to ensure all maintenance labor is optimized and utilized to the fullest extent. Spend time with maintenance schedulers. Ask, *What does your weekly scheduling process look like? Are you provided with a weekly calendar reflecting the direct labor resources who will be onsite during the scheduling window? Do the operations and maintenance frontline supervisors help when deciding what work will be on the schedule based on the available resources? Are the materials needed verified onsite?* Attend the weekly scheduling meeting to determine if the labor available matches the labor scheduled.

Daily scheduling or work assignment is typically performed by the maintenance frontline supervisor. Attend the daily scheduling/work assignment meeting if it exists to determine if labor available matches labor scheduled. If not, ask the maintenance frontline supervisor how the work is assigned to the maintenance technicians.

Figure 4.13 presents an example of a work scheduling scorecard.

Maintenance Work Management Assessment

Friday, June 25, 2021

Criteria	Pts	Score	Comments
Work Scheduling			
Weekly Scheduling -- Is the Weekly Scheduling Meeting Agenda Followed? a. Review status of any deviations from this week's schedule b. Review work to be performed next week c. Assure Planning status is "Ready to Schedule" d. Determine when to reschedule remaining work	20	5	Weekly scheduling meetings exist, and status of the current scheduling period is reviewed. Next week is scheduled and communicated to the stakeholders. This process is facilitated and communicated using Excel. No mechanism exists for work center loading and resource balancing. SAP status is not reviewed. Only building a one-week schedule, so work pushed beyond one week is not readily visible. Schedule Compliance and Break-in KPIs indicate that 50% of labor is scheduled. This was verified in field observations.
Weekly Scheduling Meeting -- Are required attendees present? a. Scheduler b. Process Area Supervisor c. Maintenance Supervisor d. Process SLS & Mechanical SLS (optional)	20	10	Scheduling is conducted by the planner and maintenance supervisors. No scheduling role exists.
Weekly Scheduling -- Are meeting attendees prepared for the meeting with the following: a. Copy of four-week schedule b. Copy of current week's schedule c. Knowledge of job planning status d. Process - discuss and justify exceptions (only) to the four-week schedule	10	2.5	No four-week schedule. The current schedule includes next week and only minimal long-term issues that have been identified. No full schedule for week T-2 and beyond is discussed.
Weekly Scheduling -- Is the weekly schedule (frozen) completed on Thursday for the next workweek?	10	10	The schedule is set after the Thursday scheduling meetings and communicated on Friday. This schedule is used as the baseline for the weekly Maintenance Metrics.
Daily Scheduling -- Is a Daily Work Assignment Meeting Agenda followed? a. Review progress of today's work assignment and identify jobs to carry over to next day b. Identify jobs/tasks that break the Weekly Schedule, but need to be scheduled for the next workday c. Identify "first start" jobs/tasks that require preparation by Operations and notify affected areas d. Discuss barriers/delays encountered during the current workday and identify possible corrective action	20	5	Currently, Daily Scheduling and Work Assignment is ad hoc and by nature covers all elements needed to progress the work. The lack of a formal process hinders the collection of data necessary to facilitate continual improvement. No formal mechanism exists for communicating the daily work assignment to Operations in preparation for job execution and permitting.
Daily Scheduling -- Is a written Daily Work Assignment form completed and posted for the next workday so craft personnel can review it?	10	0	A Daily Work Assignment Sheet is not utilized. Most work assignment is communicated through the weekly schedule, work orders, and verbal communication.
Are all base maintenance resources (company / contractor) scheduled weekly? (< 50% = 0 pts, < 75% = 5 pts)	10	0	Weekly scheduling is an ad hoc process where Operations' wants and needs are communicated. Schedule Attainment performance and Break-in Rate indicate that 50% or less of the resources are scheduled.
	100	32.5	

FIGURE 4.13 Work scheduling scorecard.

Work Execution and Close Out Scorecard

Work execution is where all previous activities can be evaluated. Every element of the work should be identified and provided for the maintenance frontline supervisor. Test the technicians to determine if they are aware of their work assignments before the start of the workday.

Travel with the maintenance frontline supervisor to observe the supervisor's daily routine. Travel with the technicians to determine the quality of the plans. Feedback from the technicians is vital for the maintenance planners and reliability engineers.

Test the entry of data the technicians are recording in the work order (CMMS) for actual time worked on each task, failure and cause codes, equipment condition, and planning and engineering feedback.

Figure 4.14 shows an example of a work execution and close out scorecard.

Maintenance Work Management Assessment

Friday, June 25, 2021

Criteria	Pts	Score	Comments
Work Execution and Close Out			
Do the Maintenance Supervisors use a Daily Work Assignment tool to make the job assignments in the morning and, are goals communicated indicating the amount of time a job/task is expected to take to complete?	10	0	Work assignment is done weekly via the weekly schedule in some areas. Daily work assignment is done verbally, ad hoc, typically in the morning with updates throughout the day as needed with changing priorities.
Are the craftsmen at the worksite with tools, parts, and materials ready to walk down the job within 45 minutes of the shift start?	10	2.5	Tools and materials are typically available, but some delays are created by break-ins and the associated vendor requests. Delays were observed due to no existing plan being available. Craftsmen are planning work in the field, determining the materials and tools needed and experiencing delay.
Is Operations aware of job preparation requirements, and has equipment been prepared for Maintenance to start work? (notification of work the night before?) Are permits prepared in advance of the job?	10	0	There is no formal process for daily work assignment, daily scheduling, and the communication of equipment and permit requirements in advance of the start of work. The reactive environment and changing priorities create a "when-needed" environment.
Are there any Permit delays? If yes, how long on average? Are permit delays being captured?	10	7.5	Field observations indicate that permit delays are minimal. Operations responds quickly to requests for permits as most work is high priority in a highly reactive environment.
Do the Maintenance Supervisors spend time in the field (at least 50% of total time) conducting positive follow-up on the work in progress addressing barriers encountered and providing recognition to the crew for good performance? (>50% = 10 pts, > 30% = 5 pts, < 30 % = 0 pts.)	10	10	Due to the reactive nature of work and the elevated break-in rate, the Maintenance supervisors are required to be in the field well over 50% of their time.
Are required parts and materials available, or do the crafts need to make a trip for additional parts? Is required transportation equipment (rigs, etc.) at the jobsite as scheduled? Is prework (e.g., scaffolding) scheduled and completed in a timely manner?	10	5	For the most part, parts and materials are available when needed. Some delays are with coordinating contractor support, i.e., scaffolding. Minimal planning and scheduling are creating a reactive environment where requirements are discovered during execution. Supervisors report that local contractor equipment and staff are utilized, are abundant in the local market, and respond quickly to the required needs. Scheduling is minimal creating a reactive environment and delayed execution.
Do the craftsmen record their actual hours worked, work completed details, failure and cause codes, and provide feedback to reliability engineering and work planning in the CMMS? Does the maintenance supervisor review the work orders completed each day and complete them in the CMMS?	10	0	At this time, the corrective work is not planned in the CMMS to the task level. Some FM work has minimal plans. Work orders are utilized as placeholders on a schedule and do not collect execution performance and reliability data.
Is there obvious lost time observed, e.g., late starts, early quits, and long breaks? Are these delays being captured?	10	5	Due to a lack of planned and scheduled work, when there is no high-priority work, the urgency is not driven by a queue of executable work. Field observations indicate that daily work assignment does not fill the available labor.
Has support craft work been coordinated so there is no lost time? (support craft scope included in the job plan and scheduled accordingly)	10	5	Support craft being coordinated as needed by the maintenance supervisors. Support craft are not included in the work orders. Observations indicate that support craft respond as defined by a break-in process.
Are break-ins given directly to the Maintenance Supervisor?	10	10	Due to the reactive nature of most work, the Maintenance Supervisors are aware of all break-ins shortly after being reported and are actively involved in the minimal planning and rapid execution of break-ins.
	100	45	

FIGURE 4.14 Work execution and close out scorecard.

Barrier Teams and Maintenance Stewardship

Schedules are only good the second they are locked. The variables that break the schedule are too many to even think about, and they increase as time passes through the duration of the schedule. That's why we develop a new schedule every week. This is no reason to get discouraged. Tracking and solving barriers to schedule attainment are a part of normal maintenance operations.

Business is business; we have to have baselines and measure progress to continually improve. Process and performance stewardship are needed at regular intervals. Attend the barrier team and maintenance stewardship meetings if they exist, to evaluate their adherence and performance to their accepted agendas.

An example of a barrier team and maintenance stewardship scorecard is shown in Figure 4.15.

Maintenance Work Management Assessment

Friday, June 25, 2021

Criteria	Pts	Score	Comments
Barrier Team & Maintenance Stewardship			
Was the last Barrier Team Meeting held within the last month?	10	0	No Barrier Team Meeting exists
Is the Operations Supervisor the Barrier Team Meeting leader?	10	0	No Barrier Team Meeting exists
Are "mandatory" team members present in the Barrier Team Meeting (Maintenance Supervisor and Operations Supervisor)?	10	0	No Barrier Team Meeting exists
Did the team follow the Agenda and use the problem-solving process?	10	0	No Barrier Team Meeting exists
Were barriers resolved or an action plan developed to address barriers?	10	0	No Barrier Team Meeting exists
Were Maintenance KPIs published this month?	10	10	Maintenance KPIs are published weekly and rolled to monthly.
Were the KPIs published weekly for the past month?	10	10	Maintenance KPIs were published weekly.
Were labor hours and Backlog numbers published for the past month?	10	10	Numbers published weekly.
Were the Daily Work Assignment KPIs published monthly for the past month?	10	0	No Daily Work Assignment process being utilized.
Was a monthly Maintenance KPI Stewardship Meeting held this month?	10	0	Weekly Maintenance KPI reports are published weekly and distributed. No meeting is held to review with Operations and Maintenance teams.
	100	30	

FIGURE 4.15 Barrier team and maintenance stewardship scorecard.

Maintenance Work Management Assessment Scorecard

This scorecard is very important to gain an understanding of your perceived as-is MWM processes and to set a baseline for processes and performance measurement of future adherence to your new, well-thought-out MWM processes that you will design. You will use this scorecard in the future to conduct periodic audits after implementing your new MWM processes.

The information provided by this assessment will be a vital piece of your business case when asking management to invest in asset and equipment reliability and maintenance.

For an example of an MWM assessment scorecard, see Figure 4.16.

Maintenance Work Management Assessment

			Friday, June 25, 2021
Work Identification and Approval	100	20	
Work Planning	100	25	
Work Scheduling	100	32.5	
Work Execution and Close Out	100	45	
Barrier Team & Maintenance Stewardship	100	30	
Total	**100**	**30.5**	

FIGURE 4.16 Maintenance work management assessment scorecard.

TIME-ON-TOOLS STUDY

Commonly called *craft productive time*, utilization is the total hours of hands-on work performed per day. It is the amount of craft labor required to perform direct work without any lost time between activities or tasks. Craft productive time can be improved by increasing the amount of work performed in the same time frame.

Lost time is measured and quantified as late starts at the beginning of the work shift, time spent not working between jobs, and finishing early at the end of the work shift. Time on tools (ToT) is also commonly called *wrench time* within industry and can be improved by increasing the amount of work performed in the same time frame.

Time-on-Tools Study Benefits

Well-planned and scheduled work provides all the resources, materials, and information needed before the start of the work shift, providing for faster repair and completion times, less lost operating time for production, and less variability in repair times and costs. Increased productive time provides more predictability in operating costs feeding the budgeting process. Productive time improvements lead to increased customer satisfaction, better improved peer relationships, and job security.

The ToT study identifies the barriers to work completion. The benefit of barrier identification is the elimination of the barriers at a systemic level through process-defined roles and responsibilities and better planning, scheduling, and work assignment.

Best Case Productive Time

As the term *time on tools* suggests, only personal contact with a tool or asset is counted as productive time. The ToT process accounts for nonproductive time that is acceptable, like travel and safety permitting. It excludes time allowed for lunch and scheduled breaks. I have seen several methods that are used to save time, but they are not statistically valid. I know this because before I trained, I challenged my trainers. They insisted, and I tested their rationale. They were correct, as the number of observations increase, the return on investment decreases.

ToT is based on observations and recordings of labor at 15-minute intervals throughout the entire workday from start to finish. The achievable maximum attainable ToT percentage increases and decreases as the workday duration increases and decreases. The longer the day, the higher the achievable percentage of productive time available. ToT allows for liberal time intervals for lost time categories and promotes continual improvement. It is not based on actual or perceived histories of past ToT studies; each study stands on its own.

Figures 4.17–4.19 provide examples of a productive time benchmark for, respectively, an 8-hour, 10-hour, and 12-hour workday.

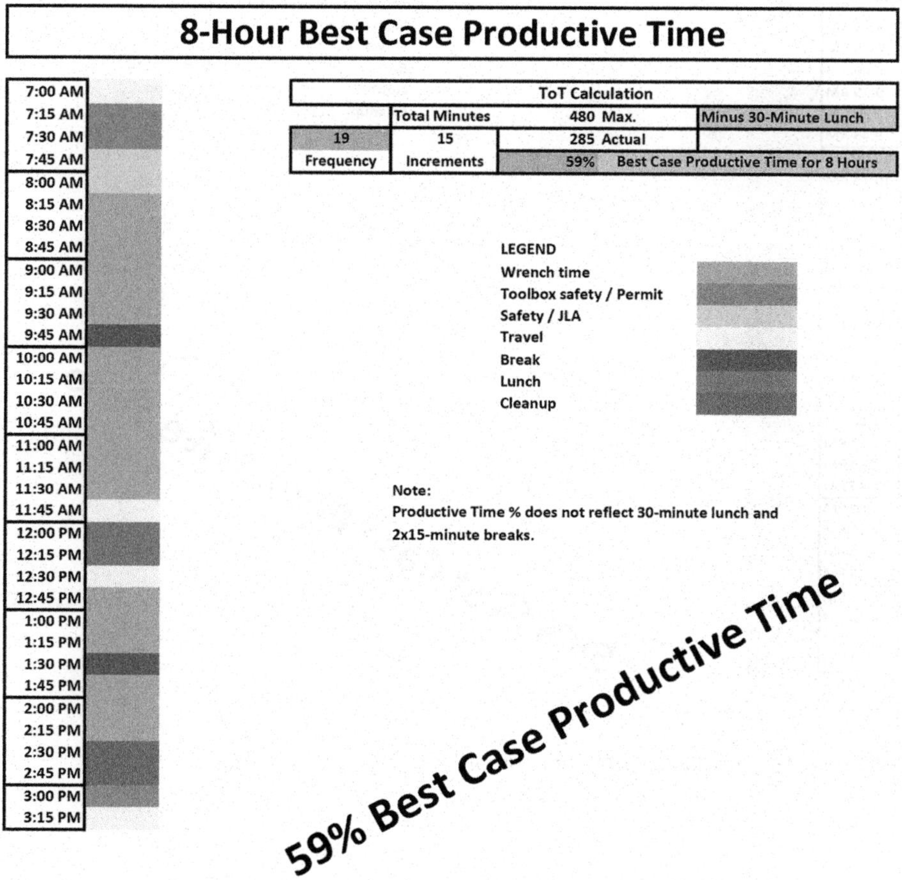

FIGURE 4.17 Productive time benchmark for an 8-hour workday.

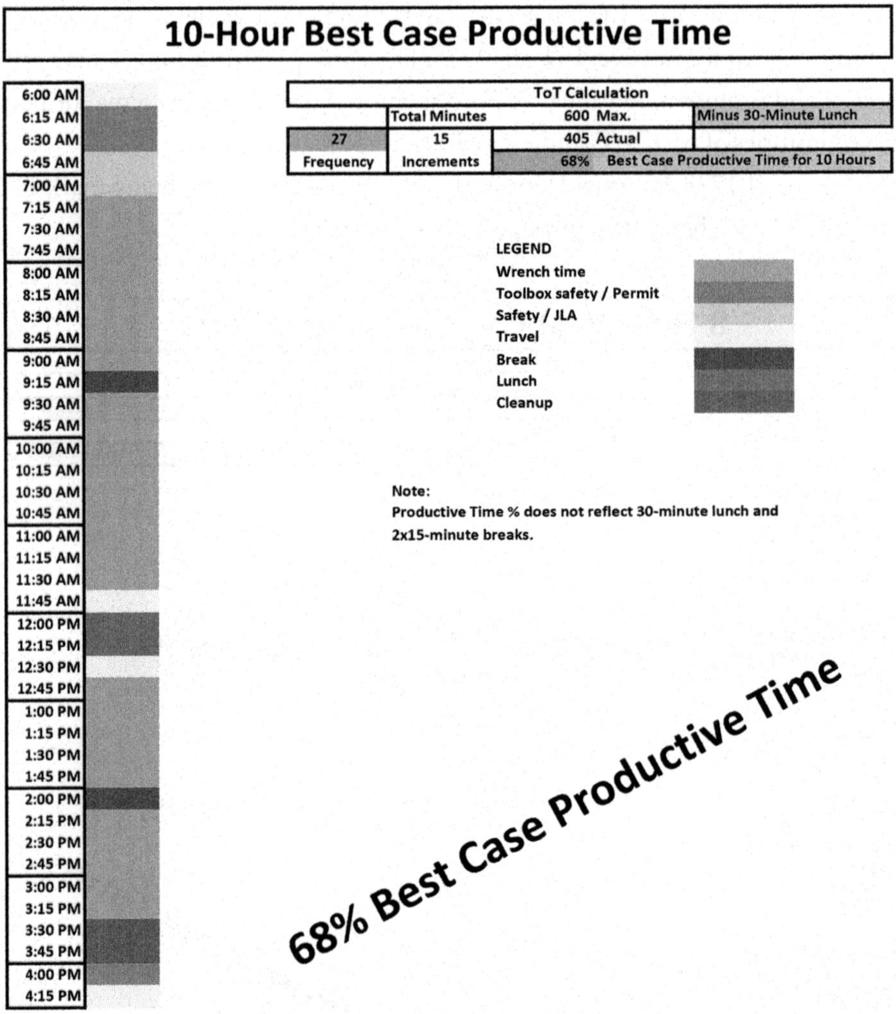

FIGURE 4.18 Productive time benchmark for a 10-hour workday.

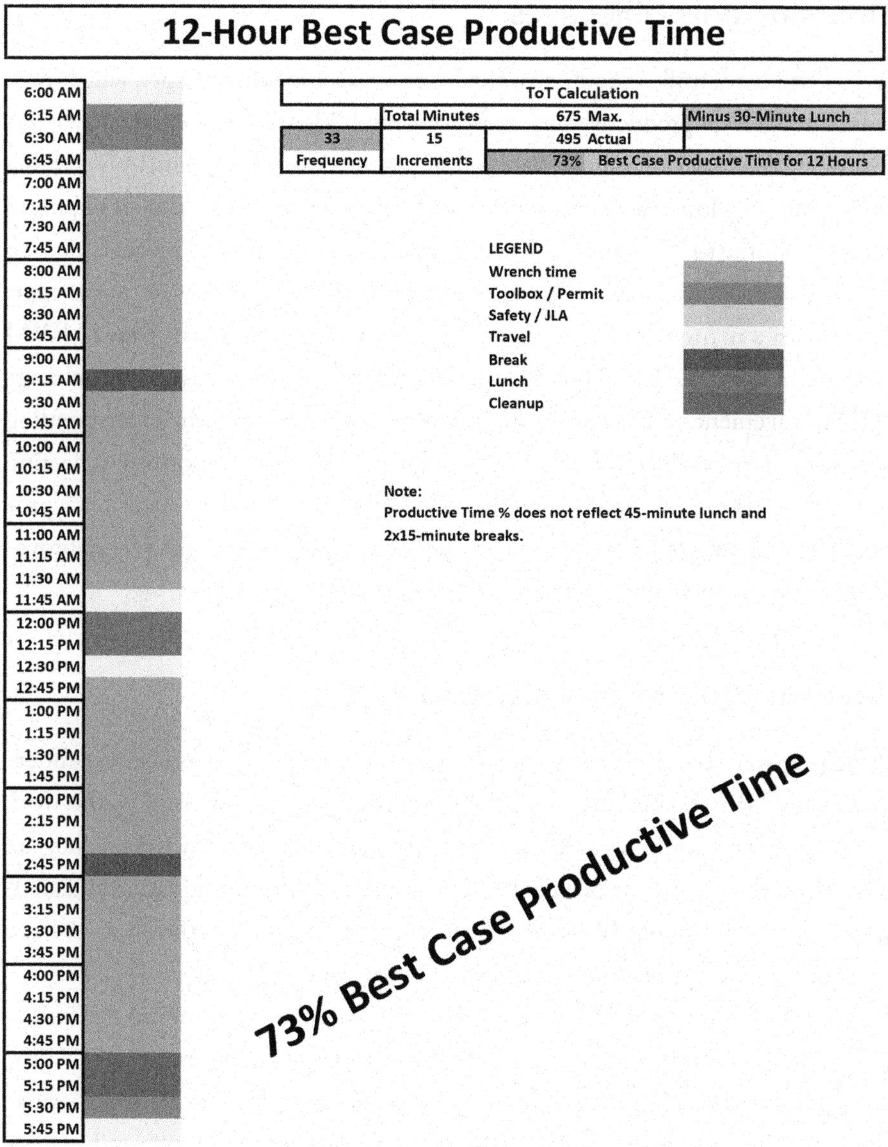

FIGURE 4.19 Productive time benchmark for a 12-hour workday.

Time-on-Tools Objectives

A ToT study identifies productive time versus lost time by category, which provides direction for productive time improvement. It identifies successes and gaps in maintenance work assignment, field follow-up, and work execution, by qualifying and quantifying lost-time opportunities, leading to recommendations for improved productive time to the maintenance work management (MWM) processes.

The ToT study provides documentation, analysis, and review to be combined with a maintenance work management scorecard, a brown paper MWM gap analysis, and DILO stakeholder studies to build a business case for change and improvement. A quantified productive time increase and the corresponding cost savings provide a dollarized benefit for management's approval, reflecting improvements to performance and the MWM processes. It also establishes a productivity benchmark for future ToT studies and can be used to measure improvement after a change in performance or the MWM processes.

Time-on-Tools Methods and Tools

Well-prepared ToT observers come armed with the MWM processes' understanding of the stakeholders' roles and responsibilities. They observe the work orders and plans of the work to be executed to understand the activities they are observing. They make use of a data entry form and a data consolidation and analysis tool, typically a standard spreadsheet format. It is helpful if the observer has a basic understanding of common spreadsheet software.

The ToT observer follows a crew from the work shift beginning through to the completion of the day, identifying productive work, away time, and idle time and recording the 15-minute observations on the data entry form. The information from the data entry form is transferred to the data consolidation and analysis tool to produce charts that are then analyzed and used for reporting.

Observers are objective during the ToT observations and analysis, and do not actively coach while in the field, avoiding influence on the results of the study. They schedule management reviews to present their final report and recommendations for improvement, giving feedback and coaching opportunities to management.

During the observations, the observers are assessing to the MWM process, preparing recommendations, and reporting the efficiencies of the work crews' behavior. The observers consider the crews' adherence to the processes to ensure that all crews participate in morning line-up sessions, that job assignments are complete, and that crews move to the jobsite expeditiously.

They evaluate the timeliness of the permit issuance, identify lost time due to excessive travel, review job plans with supervisors prior to work observation so the observers understand the work that is to be done, observe work in the field to ensure supervisors are following job plans and packages, assess the overall quality of field execution to include contractor coordination, and observe the frequency and quality of field follow-up by supervision.

Time-on-Tools Process

Observers identify the area to be observed, are in the area prior to the start of the work shift, and pick the crew to observe, not allowing for supervisor preference. Crews must vary by discipline over the duration of the entire study for accurate averaging. Different types of work are more complex, and following only complex work distorts the observation to a positive result as these jobs are longer in duration. At the other end of the spectrum, short-interval jobs like inspection can distort the results to a negative outcome, as the crew may spend large amounts of time consuming acceptable away time such as travel.

Observers stay with the work crew from the beginning to the end of the work shift unless the crew disbands. If the work crew separates and performs more work, continue observation with one of these crews until the end of the shift. Inquire where crew members have gone if they leave the worksite, and record their observation based on the best information received. Do not use a clipboard or notepad openly, as this can be unnerving to the work crew; instead record observations discretely. I like to print a copy of the data collection form and keep it in my pocket, recording observations when I can't be directly observed.

Record a minimum of 800 data points over the entire ToT analysis for statistical validity. Observe the activity of the individual crew members in 15-minute intervals. One crew member times 15 minutes equals one data point.

Example: 3 observers, 10-hour shifts for 3 days

- Total data points 1 craft per 10-hour day = 40
- Average crew size 3 crafts = 120 data points per day
- 3 observers = 360 data points per day
- 3 observation days = 1,080 data points

After performing more ToT studies than I can remember, I can safely say that the 800 or more data points needed for statistical validity are achieved in 8 man-days.

Time-on-Tools Categories

These are categories that are found in data consolidation form spreadsheets. To begin, understand the definition of *productive work*:

> *Productive work.* Performing work that is planned and requires maintenance direct labor resources.

Next, understand the meaning of *away* and *idle*:

Away. Time that is consumed away from the worksite:

- **TRAVEL.** Crew traveling to and from jobsite
- **UNSCHEDULED PERSONAL BREAKS.** In addition to the normal schedule
- **PARTS/TOOLS/EQUIPMENT.** Waiting at stores or leaving site to obtain
- **WORK ASSIGNMENT/TOOLBOX SAFETY.** Required and scheduled
- **PERMITTING/SIGN-IN.** Operations requirement
- **EXTENDED BREAKS.** Taking longer than scheduled
- **MEETINGS.** Other than work assignment and safety
- **WEATHER**

Idle. Time at the worksite not working:

- **OPERATIONS DELAY.** Includes equipment not available
- **WAITING FOR RESOURCE NOT IN PLAN.** Caused by planner
- **WAITING FOR RESOURCE IN THE PLAN.** Support craft not available
- **WAITING FOR INFORMATION.** Engineering specs, equipment specs, technical data, etc.
- **REWORK.** Redoing work on the same work order or asset
- **WAITING FOR PARTS, TOOLS, EQUIPMENT.** At jobsite, waiting for others
- **WAITING FOR NEXT ASSIGNMENT.** Supervisor assigning next work
- **WAITING FOR DIRECTION.** From supervisor or subject-matter expert
- **BREAK-INS.** Time lost when reassigned
- **FIELD PLANNING/SCOPING.** Inadequate plan or no existing plan
- **PERMITTING/JOB SAFETY ANALYSIS.** In excess of allotted time

Helpful observation points include:

- Get a copy of the work order and plan.
- Observed crew comfort matters.
- Do not interfere with the work being done.
- Record observations outside the line of sight.
- Create a friendly atmosphere by asking for help.
- Ask their first name.
- Explain you are there to "see what stops them from working."
- Do not write things down in front of them.
- Never disagree or try to change their course of action.
- When time permits, ask them what stops them on a typical day.
- Listen, listen, listen.
- Thank them, thank them, thank them.
- Respect every individual.

Time-on-Tools Data Entry Tools

In my many years of conducting ToT studies, I have not come across a ToT data entry and reporting app or software. This is typically handled with a standard spreadsheet tool. I did come across a tool from Europe that counted an entire crew regardless of size. This reduces the number of actual observations and skews the consolidated averages, and thus it is not statistically valid.

With this spreadsheet tool, readings are recorded every 15 minutes on individual lines, and one numbered column across the top represents one crew member. Figure 4.20 shows an example.

As mentioned earlier, as crew size increases, keeping track of all crew members can become a challenge. If crew members leave the worksite, ask them where they are going, and record the observations every 15 minutes until they return. If you notice a crew member missing, ask one of the members of the team where the crew member went and the purpose of the absence, and record this in your observations until the crew member returns.

Advanced Management, Inc.

PRODUCTIVE LABOR UTILIZATION

1=PRODUCTIVE | 1=Idle | 1=AWAY

Foreman: Janis Doe
Department: Maintenance - Mechanical
SHIFT: Day
Group Size: 3

DATE: 14-Apr-19
OBSERVATION: 6:00 AM TO: 6:00 PM
1st BREAK TO:
LUNCH: 11:30 AM TO: 12:30 PM
2nd BREAK: TO:

Work Permit: 101066
Task: 22609123 Crane PM
11115654 Crane Repair

TIME	1	2	3	4	5	6	7	8	9	10	SUPER. PRES.	SUPERVISORS REACTION/COMMENTS	Travel Time	Unscheduling Personal Breaks	Returning parts/tools /equipment	Work Assignment / Tool Box Safety	Permitting /Sign-in	Extended Breaks	Meetings
# 6:00 AM																			
	1	1	1									Waiting for tool box meeting to start				3			
# 6:15 AM																			
	1	1	1									Tool box/safety meeting				3			
# 6:30 AM																			
	1	1	1									Tool box/safety meeting				3			
# 6:45 AM																			
	1	1	1									Tool box/safety meeting				3			
# 7:00 AM																			
	1	1	1									Travel time	3						
												%							
TOTAL	44	43	44	1	0	0	0	0	0	0	132		6	0	0	12	0	0	0
PROD. WORK	24	22	22	0	0	0	0	0	0	0	68	52%							
ADD'L OPPTY	13	14	15	1	0	0	0	0	0	0	43	33%							
AWAY	7	7	7	0	0	0	0	0	0	0	21	16%							

Observer:

(Columns "Travel Time" through "Meetings" are grouped under the heading "Away".)

FIGURE 4.20 ToT data entry form spreadsheet. The section on the right calculates totals from entries on the left for the Away and Idle categories.

Outputs from the Data Entry Tool

The consolidation form spreadsheet shows the individual observations made for the entire ToT study (see Figure 4.21). Typically, a set of pie charts reflecting the results of the entire study are a good indicator of conformance to the maintenance work management processes and can be used to build a business case for investing in well-defined maintenance work management processes.

I have encountered management that would like to see more granularity in the data. In a spreadsheet format, the data can be easily sorted and separated to meet your needs. I have published outputs as requested for the overall observation location, employees versus contractors, individual area within a location, craft discipline, date, supervisor, and shifts.

To build your business case, use the data results from the consolidated data entry tool to suggest quantifiable improvements. Expect that on average, in a mature maintenance operation that does not have well-defined MWM processes and a solid planning and scheduling function, the results of the ToT study may be a little shocking.

Most industry sources indicate an average of 35% ToT, while I have typically found 35% in small organizations because they are more agile and the location is smaller. In very large organizations on a large geographical footprint, I have found ToT results as low as 19%. Don't be discouraged when arriving at these results; raising 35% to 52% and raising 19% to 40% show a financial gain of 50% achieved by completing more work within the same time frame that can be dollarized using an average hourly rate.

		Away							Idle									
		Travel Time	Unscheduling Personal Breaks	Returning parts/tools/equipm	Work Assignment / Tool Box Safety	Permitting / Sign-in	Extended Breaks	Meetings	Operations Delay	Waiting on resource not in Plan	Waiting on resource in the Plan	Waiting on information	Rework	Waiting Parts, Tools, Equipment	Waiting for Next Assignment	Waiting for Direction	Break-ins	Field Planning / Scoping
LNG MECH																		
PROD. WORK	52%																	
IDLE	33%	6	0	0	12	0	0	0	9	0	0	0	0	6	17	11	0	0
AWAY	16%	18							43									
		33%	0%	0%	67%	0%	0%	0%	21%	0%	0%	0%	0%	14%	40%	26%	0%	0%
LNG SCAF								100%										100%
PROD. WORK	44%																	
IDLE	7%	29	0	59	27	2	0	0	0	0	0	0	17	0	0	0	0	0
AWAY	49%	117							17									
		25%	0%	50%	23%	2%	0%	0%	0%	0%	0%	0%	100%	0%	0%	0%	0%	0%
LNG ELE								100%										100%
PROD. WORK	50%																	
IDLE	35%	3	0	0	0	0	3	9	3	0	0	0	0	12	12	0	0	5
AWAY	15%	15							32									
		20%	0%	0%	0%	0%	20%	60%	9%	0%	0%	0%	0%	38%	38%	0%	0%	16%
HGCP ELE								100%										100%
PROD. WORK	24%																	
IDLE	17%	8	0	10	3	2	0	2	0	4	0	0	0	0	0	0	0	3
AWAY	60%	25							7									
		32%	0%	40%	12%	8%	0%	8%	0%	57%	0%	0%	0%	0%	0%	0%	0%	43%
LNG INST								100%										100%
PROD. WORK	42%																	
IDLE	4%	20	0	9	12	28	0	3	0	0	0	0	0	0	6	0	0	0
AWAY	53%	72							6									
		28%	0%	13%	17%	39%	0%	4%	0%	0%	0%	0%	0%	0%	100%	0%	0%	0%
LNG ROPE								100%										100%
PROD. WORK	61%																	
IDLE	18%	31	0	0	6	0	4	0	20	0	0	0	0	10	10	0	0	0
AWAY	21%																	
		34%	0%	27%	21%	11%	2%	5%	23%		0%	0%	12%	20%	32%	8%		6%
PROD. WORK	48%							100%										100%
IDLE	18%																	
AWAY	34%	288							141									

FIGURE 4.21 Sample data consolidation form spreadsheet. This information is pulled and consolidated from the data entry form.

Time-on-Tools Results

Figures 4.22–4.26 show the results from a very large operation spread over a large geographical area that was only seven years old. This operation was in a remote location requiring onsite accommodations for management and labor resources working 12-hour days on a rotational schedule.

When evaluated, a fair percentage of the away time and 100% of the idle time could be eliminated. The goal for improvements in this case reflects an improvement of 49% to 73% in the best case. This reflects a 49% opportunity for an increase in productive time.

When presenting results to management, and I agree as a manager myself, the easier to read, the better to save time. Even more so when the color reflects good and bad results. I think we all learn at an early age that red means stop because bad things can happen, green means proceed because moving forward is good, and yellow means take action. Time is limited for management—make their job easier.

FIGURE 4.22 Total ToT observations.

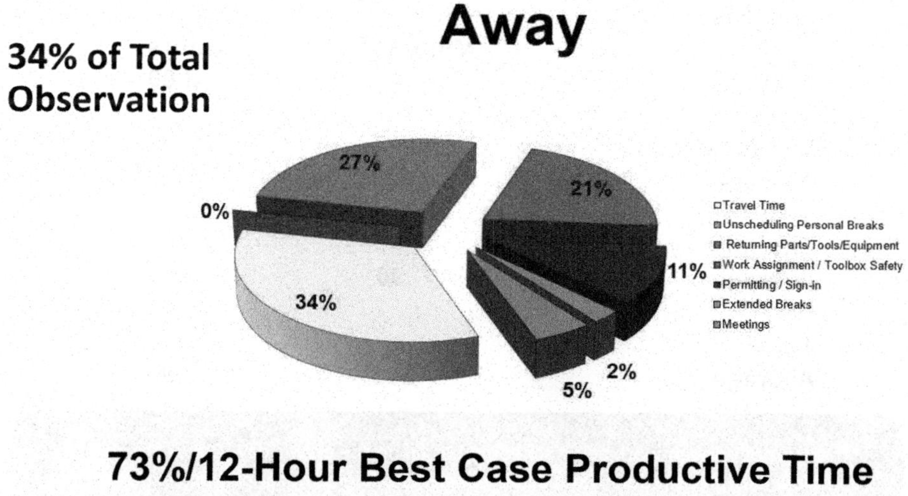

FIGURE 4.23 ToT away time.

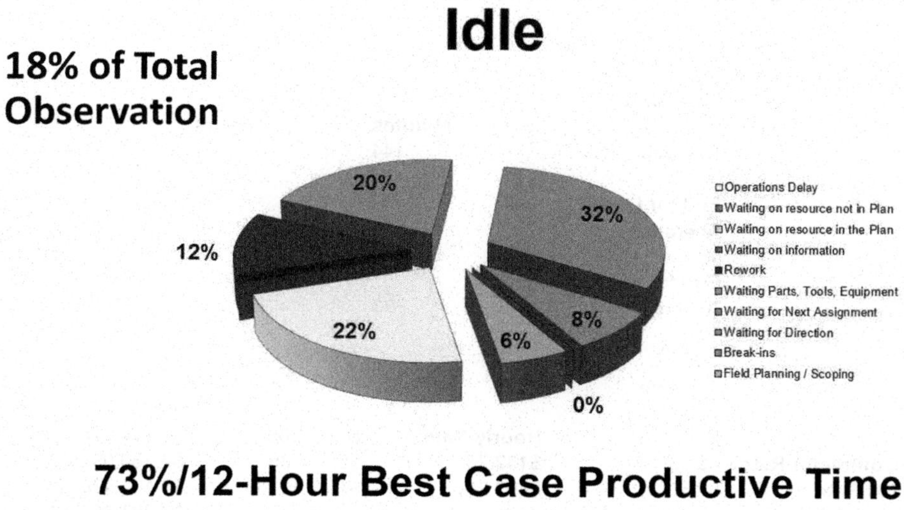

FIGURE 4.24 ToT idle time.

First Start	Minutes	Hours
Average	144	2.00
End for Lunch		
Average	29	0.4
End for Day		
Average	*90*	*1.5*
Total		
Average	263	3.90

There is not enough work being scheduled to fill up the crew's day. Crews are able to complete existing schedules (and achieve high schedule compliance while experiencing break-in rates of 9-13%) and still have the capacity to do more. That excess capacity manifests itself in late starts and early quits for lunch and the end of the shift.

FIGURE 4.25 Daily recurring lost time—easily eliminated.

	First Start			
			Minutes	Hours
	Average		144	2.00
	End for Lunch			
	Average		29	0.4
	End for Day			
	Average		*90*	*1.5*
	Total			
	Average		263	3.90

	FTE	Hourly Rate	Daily Cost	Yearly Cost 1 Hr
Contractor Rate	45.00	$132.73	$5,972.85	$2,180,090.25
EM Rate	63.50	$263.25	$16,716.38	$6,101,476.88
				$8,281,567.13
	Yearly Cost 1.95 Hr		$16,149,055.89	

FIGURE 4.26 Direct lost time—start late/end early. A 49% ToT increase creates $16 million in potential savings on inefficiencies and lost time.

MWM BUSINESS CASE FOR CHANGE

Using the brown paper MWM as-is processes analysis, the DILO studies, the MWM scorecard, and the ToT study, a powerful case can be made for designing, implementing, and sustaining new MWM processes. Management will appreciate your report details, qualifications, quantifications, savings, and details for moving forward with your project and savings.

Project Objectives and Path Forward

- Increase productive time from 49% to 64%
- First year's savings potential of $4.6 million
- Potential annualized savings of $7.9 million
- Consistent maintenance work processes, tools, and practices
- Increase productive time by two hours per craft shift (from six to eight hours)

Work Process Design and Documentation—10 Calendar Weeks

- Work notifications
- Risk-based work selection
- Planning and scheduling
- Daily maintenance work execution
- Barrier teams and KPI stewardship

MWM Implementation—Tools, KPIs, Skills Training and Behavioral Coaching—20 Calendar Weeks

- Operators—work notifications
- Ops supervision—risk-based work selection
- Planners, schedulers, and material coordinator—planning and scheduling training
- Maintenance execution supervisors and craft leads—execution
- Contractor supervisors and leads—execution
- Supervisors and superintendents—work process and KPI stewardship

Project Cost

If you're using this book to conduct your project using internal resources, your cost will be absorbed by the cost of your involved stakeholders in time dedicated to the project in the normal course of their employment. This project will make their jobs more efficient and easier. If you intend to use external resources, it will be helpful to present that cost to management as part of your business case.

CHAPTER FIVE

DESIGNING THE PROCESS

MAINTENANCE WORK MANAGEMENT

As stated at the beginning of this book, maintenance work management is the structured process of identifying, prioritizing, planning, scheduling, executing, documenting, closing, and analyzing maintenance tasks such as repairs, inspections, projects, upgrades, and preventive care, to optimize equipment reliability and minimize downtime. Utilizing tools like computerized maintenance management systems (CMMSs) ensures improved safety, efficient resource allocation, cost control, and reliability and extends the life of physical assets.

Key Principles

> *Planned work is safer and more cost-effective. The safe execution of work is accounted for in the individual work plans.*

The members of the field operations team are responsible for the daily operating and monitoring of the assets. They are viewed as the owners of the assets and maintenance budget. The maintenance budget is an element of future reliability investment and cost, as an element of the overall return on investment.

The members of the reliability and maintenance team identify, develop, and execute programs to increase equipment reliability and production uptime in a safe and cost-effective manner. They promote reliability-based maintenance best practices to minimize equipment breakdowns and unplanned deferred production events while controlling maintenance costs to achieve maximum net return.

MAINTENANCE WORK MANAGEMENT PROCESS DESIGN

The new maintenance work management (MWM) processes must be customer-focused, contain no non-value-added steps, and be free of redundancies and bottlenecks. Be very deliberate defining each step of your processes and the questions that need to be answered before a notification or work order moves to the next step. When possible, identify steps in parallel when responsibilities are shared between functions, such as the maintenance frontline supervisor observing the technicians' execution of work and the operations supervisor accepting completed work quality and documentation with the maintenance technician. Create smooth hand-offs between the six MWM processes and the stakeholders responsible for them. It is very helpful at this point if one of your design team members has and knows how to use a well-known process flowcharting software.

Define clear roles and responsibilities for each functional position. Define a matrix of the skills necessary to perform each of the functional positions and the needed training. The stakeholders must be capable and trained to utilize the established workflow, and the workflow must be supported by a user-friendly information system whether it's a CMMS or manual system. Management must provide a supportive culture that promotes the success of the MWM process design, implementation, and sustainability.

Process Management

Key stakeholders are selected to take responsibility for process ownership, planning, and resource allocation and ensure inter-process links with other functions in the organization. Continual improvement is adopted, and upstream process metrics are developed to record process monitoring and performance using an easy-to-understand reporting system.

Maintenance Work Management Process

When moving to maintenance, after many years in operations, I learned that MWM follows the same natural processes as production operations. In operations, we forecasted sales or took customer orders, considered our understanding of the customer need and fit for our processes and capabilities, planned the details to deliver our product or service, considered our capacity based on demand, scheduled our resources, built the product or delivered the service, and collected the resulting data and analyzed our performance for continual improvement. Just like MWM.

Maintenance Work Management Design Elements

Elements include:

- **PROCESS FLOWCHART.** Graphically defines the process steps and sequences using standard symbols as identified with a standard, industry-accepted process flowcharting software
- **PROCESS STEP DEFINITIONS.** A list of the steps on the flowchart including the step number and description title, as on the flowchart, and the definition describing the activity taking place in that step
- **PROCESS RACI CHART.** A grid-style chart with a list of the steps on the flowchart, the organization's functional position titles across the top, and a designation in the corresponding line and column box indicating the functional position titles' interactions within the process:
  - **R** = Responsible, "The Doer"—performs the work
  - **A** = Accountable, "The Buck Stops Here"—oversees "The Doer"
  - **C** = Consult, "Before a Decision"—is asked for advice
  - **I** = Inform, "Keep in the Loop"—receives needed information

Keep in mind that the following example flowcharts, step definitions, and RACIs fit a particular organization—its management structure, product, available human resources, and culture—following as closely as possible the six standard MWM processes, and are reflected in most well-known CMMS software.

WORK IDENTIFICATION PROCESS

Maintenance work is identified using two main categories: preventive work and corrective work. The preventive work consists of any planned and repetitive work that will be completed on a set frequency utilizing preventive maintenance (PM) or predictive maintenance (PdM) equipment strategies to prolong asset life and reliability.

The corrective work is an asset found to be defective or malfunctioning needing repair. This work is not repetitive and is typically due to equipment breakdown or failure of the equipment to meet its intended purpose. Although not repetitive, job plans can be structured and put in a library for similar work to save planning time on similar types of equipment, for future use. Other categories of work performed by the maintenance resources can include project work, equipment upgrades, shutdown/turnaround work, and asset/equipment upgrades. The work identification process is the time to ask the question, *Does the needed work meet the definition of maintenance by operations, as mentioned earlier, and can it bypass the maintenance work management process?*

In most organizations, maintenance work is managed using a CMMS. In smaller organizations with no dedicated CMMS, the work can be managed with production's enterprise resource planning (ERP) system, using its work order management process by adding work centers to identify the various maintenance labor and equipment resources. One advantage of this practice is the ability to include maintenance activities on the production schedule using the ERP system. This is not to say that smaller organizations cannot use the principles in this book to develop a manual system.

In my experience, ownership and responsibility for the work identification process is with operations as opposed to maintenance because operations is responsible for the means of production, its criticality, its priority to the business. and the operating budget.

Typically, I see that 75% of maintenance corrective work is initiated by operations and 25% is initiated by maintenance technicians and others outside of operations, e.g., projects, upgrades, and shutdown/turnaround. With a high level of operations involvement in notification or work order initiation, operations supervisors and managers coach the initiators to increase quality when it is below the established standard.

Process Design Critical Elements

Elements include:

- **NOTIFICATION QUALITY STANDARD.** *What information is needed in the notification/work order?*
- **URGENT WORK.** *Does the likelihood and impact of the failure warrant immediate action, thereby bypassing the planning and scheduling processes?*
- **MANAGEMENT OF CHANGE.** *Does the work require an approved and documented change order?*
- **PM/PDM FOLLOW-UP.** *Is the work the result of preventive maintenance?*
- **NOTIFICATION GENERATION.** *Is the notification completed to the quality standard before progressing?*

Figure 5.1 is a flowchart that traces the steps in the work identification process, and Figure 5.2 defines the steps.

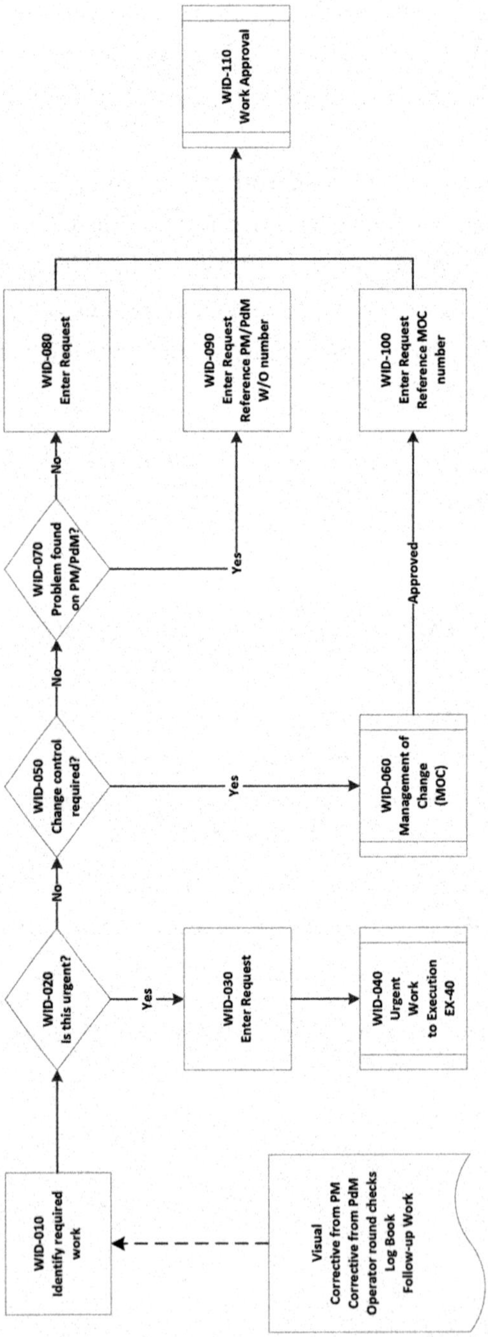

FIGURE 5.1 Work identification process process flowchart.

Process Title: **Work Identification**
Revision: **1.0**

Step Number	Step Description	Step Definition
WID-010	Identify required work	Any task requiring maintenance or repair for Operational Reliability and/or Safety/Environmental consideration. These tasks may be identified by any personnel onsite as a result of visual inspection, operator rounds, PM/PdM tasks or follow-up work. NOTE: If a worker does not have access, their supervisor should be contacted to create the notification.
WID-020	Is this urgent?	Urgency is determined based on the Risk/Priority. *Urgent: an identified deficiency that has the potential to result in an unsafe condition, health hazard, hazard to the environment or damage to onsite physical assets, or presents an unacceptable potential loss to process.*
WID-030	Enter Request	Choose correct equipment. Describe problem in detail. Choose preliminary priority. Identify possible solution. Identify parts, if known.
WID-040	Urgent Work to Execution EX-40	Work deemed urgent can proceed to the Execution process; refer to Execution WEX-40.
WID-050	Change control required?	If change control is required, refer to MOC process.
WID-060	Management of Change (MOC)	Work requiring change control can proceed to change control process. If rejected, Initiator will be notified through the MOC process. If Approved, a notification will be entered.
WID-070	Problem found on PM/PdM?	Was the problem detected during the execution of a PM/PdM work order?
WID-080	Enter Notification	Choose correct equipment. Describe problem in detail. Choose preliminary priority. Identify possible solution. Identify parts, if known.
WID-090	Enter Notification Reference PM/PdM W/O number	Reference PM/PdM work order number. Choose correct equipment. Describe problem in detail. Choose preliminary priority. Identify possible solution. Identify parts, if known. Reference PM W/O number.
WID-100	Enter Notification Reference MOC number.	Reference the MOC number in the notification. Choose correct equipment. Describe problem in detail. Choose preliminary priority. Identify proposed solution. Identify parts, if known.
WID-110	Work Approval	Refer to Work Approval process.

FIGURE 5.2 Definitions of steps in the work identification process.

The RACI chart in Figure 5.3 uses the information from Figures 5.1 and 5.2.

		Operations						Maintenance								RE	MRO		
RACI Chart for:	Work Identification	Operations Director	Superintendent	Production Area Lead	Production Area Coordinator	Gatekeeper	Shift Team Leader	Operator	Engineering and Maintenance Manager	Maintenance Manager	Maintenance Lead	Technician	Maintenance Planner	Maintenance Scheduler	Initiator	Reliability Engineer	Maintenance Buyer	Stockroom Clerk	Contractor
Step Number	**Step Description**																		
WID-010	Identify required work					A							C		R				
WID-020	Is this urgent?					A							C		R				
WID-030	Enter Request					A									R				
WID-040	Urgent Work to Execution EX-40					A									R				
WID-050	Change control required?					A							C		R				
WID-060	Management of Change (MOC)																		
WID-070	Problem found on PM/PdM?					A							C		R				
WID-080	Enter Notification					A									R				
WID-090	Enter Notification Reference PM/PdM W/O number.					A									R				
WID-100	Enter Notification Reference MOC number.					A									R				
WID-110	Work Approval																		
	R = Responsible "The Doer" A = Accountable "The Buck Stops Here" C = Consult "Before a Decision" I = Inform "Keep in the Loop"	**Revision :** 1.0																	

FIGURE 5.3 Work identification process RACI chart

Maintenance Work Notification

Tasks requiring maintenance resources are identified using a work notification. A work notification is the starting point for work clarification, data collection, and communications. This work is identified by any work initiator (person) onsite. If a work initiator does not have access to the CMMS, they contact someone with access to create the work notification. Some organizations prefer to skip the CMMS notification functionality and generate work orders at this stage in the process.

Urgency for completion of the work is determined based on a priority matrix or tool. This is reflected in the CMMS using the asset criticality, priority codes, and required end dates. Work is determined to be urgent if the potential exists to result in an unsafe condition, a health hazard, a hazard to the environment, damage to physical assets, or an unacceptable loss in the production process.

Importance of the Maintenance Work Notification

The maintenance work notification is a very important communication. Its content and quality should be such that verbal communications to clarify the notification information are kept to a minimum, freeing time spent downstream by other process stakeholders seeking clarification. At a minimum, the notification specifies the asset that needs work and the location of the asset, provides a clear definition of the defect or scope, defines the priority or urgency of the work, and identifies the originator and the originator's contact information.

In more depth, the maintenance work notification should provide the location, the asset/equipment number, the accounting information, a detailed description of the failure or scope of work, the known history of past failures, the priority or urgency, a justification for urgent work (schedule break-in), the originator, the originator's contact information, the required end date, and the appropriate level of approval.

Maintenance and Reliability Codes (CMMS)

These codes include the following:

Status Codes

Each CMMS may have its own unique status codes, but the principles are the same including entered, waiting on approval, approved, waiting planning, planned, waiting on material, engineering or drawings, scheduled, in process, rework, executed, complete, and closed. These statuses are used to designate and track where particular work orders are within the maintenance work management flow and to measure backlog in each status to maintain the balance of the process and identify bottlenecks.

Object Part/Equipment-Type Codes and Functional Location

The object part or equipment-type codes and the asset functional locations specify the individual, maintainable assets location for damage repair or maintenance activity, e.g., pump, compressor, tank, pipe, and the areas within the plant where assets are located, facilitating the locating of the assets by downstream stakeholders and detailed failure analysis by the reliability stakeholders.

Work-Type Codes

These codes indicate the type of work needed and are used for various reasons to include setting work execution priority and the sorting of completion and reliability data, e.g., CM (corrective maintenance), PM (preventive maintenance), PdM (predictive maintenance), calibration, project work, improvement, cleaning, OSHA, safety, quality.

Damage/Failure Codes

Damage and failure codes are used to identify, document, classify, and analyze asset/equipment failures, such as electrical failures, broken parts, bearing vibration, and leaks to facilitate detailed failure analysis by the reliability stakeholders and alert downstream stakeholders like planners and execution supervisors of the nature of the work to be performed.

Cause Codes

Cause codes identify, as best can be determined without detailed analysis at this point, the cause of equipment/asset damage, failure, or malfunction specific to the object part/equipment code and damage/failure code, e.g., age, lack of lubrication, overspeed, and facilitating detailed failure analysis.

Work Notification Quality Standard

The work notification conveys a substantial amount of information and is used by many functions in the organization, e.g., accounting, budgeting, management, maintenance, operations, and reliability. The quality standard can be as easy as a snapshot of your particular CMMS notification screen identifying your desired requirements. Notification quality should target a notification containing enough information to eliminate the need for verbal communications. Pay particular attention to the long description and the use of a standard template if possible. See Figure 5.4.

Enter the order description on the central header to include only the common name of equipment and the failure. The number of characters are limited so keep the description concise.

Select the PMActType that best fits activity, hit Enter; the short text will populate in both header and operation areas.

Change to: Backwards

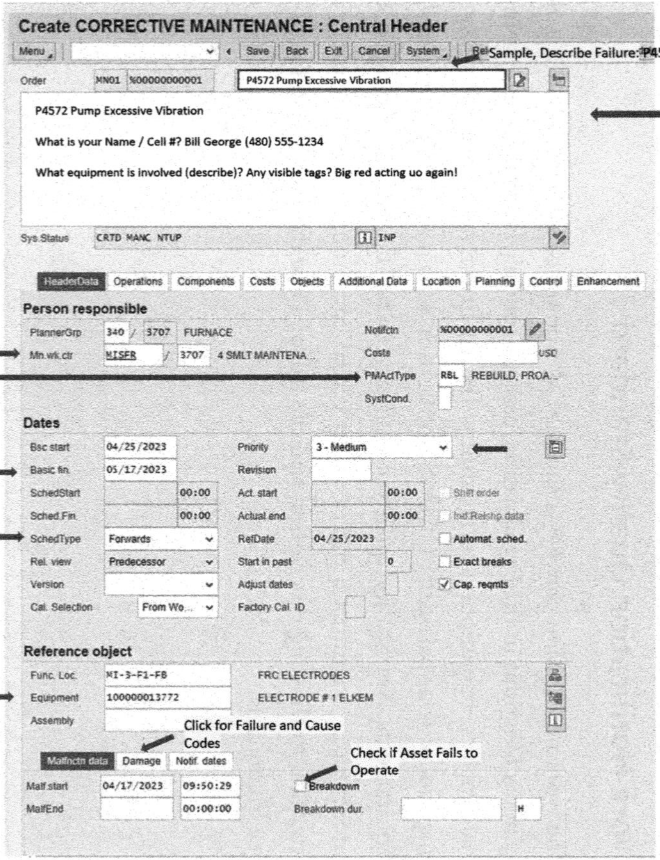

Sample Long Text Template: Describe Failure:

P4572 Pump Excessive Vibration

What is your Name / Cell #?

What equipment is involved (describe)? Any visible tags?

What do you see, hear, smell (describe in detail)?

Did you do anything to mitigate the situation?

Have you seen this before? If yes, how long was it down the
last time (If you remember when, please include appx date)?

Will this immediately impact Safety, Environmental, health, quality or production?

Did you enter Basic Finish Date to reflect Priority?

Select the most appropriate <u>Failure Mech</u> code, using the arrows to expand the options.

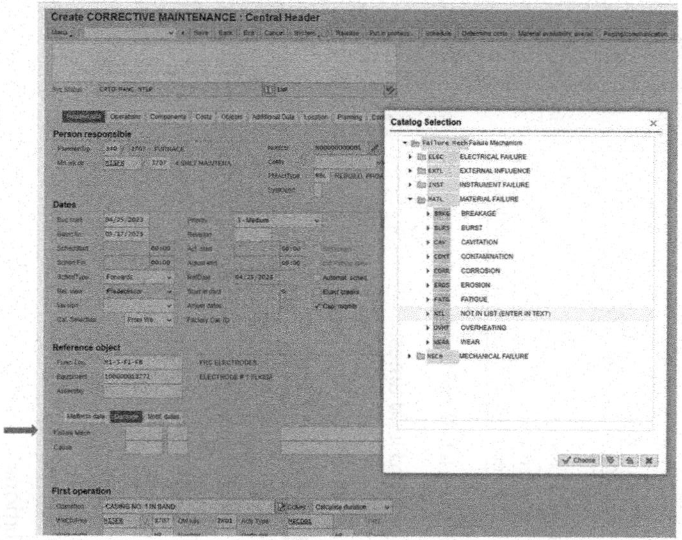

Select the most appropriate <u>Cause</u> code, using the arrows to expand the options.

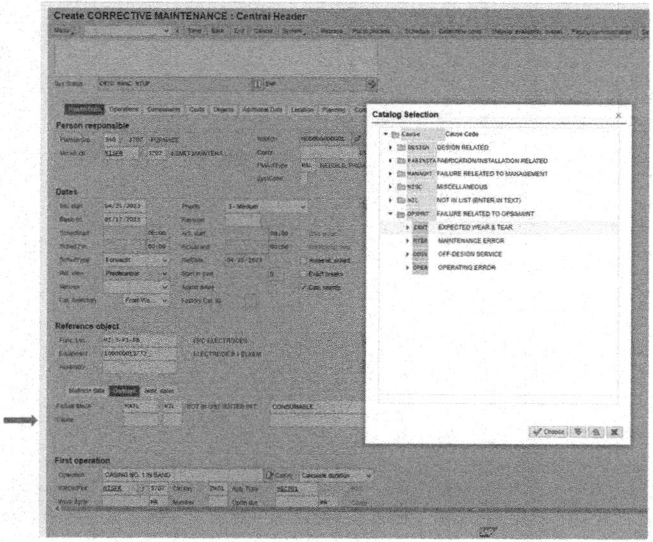

FIGURE 5.4 Work notification quality standard. (CMMS) (SAP® PM Screenshot)

Setting Work Order Priority/Required End Date/Target Finish Date

Before setting risk-based priorities and dates by using approval authority in the work approval process, the work initiator can set these informally based on their level of experience using a standard, acceptable set of criteria. Figure 5.5. shows an example for priorities and standard lead times.

Priority		Risk	Target Finish Date
Priority 1	EMERGENCY	Personal Injury	Now + 24 Hours
		Damage to Key Equipment	
		Environmental Event	
		Regulatory Event	
		Significant Production Loss	
Priority 2	HIGH	Potential to Become P1	Today + 7 Days
		Weekly Schedule Breaker	
Priority 3	MEDIUM	No Need for Weekly Schedule Breaker	Today + 8 to 30 Days
		Time to Plan	
		Perform During Scheduled Shutdown	
Priority 4	LOW	Must Have Materials	Today + 31 to 60 Days
		Must Have Resources	
		Must Have Operational Opportunity	
Priority 5	FUTURE OPPORTUNITY	Fully Planned	Today + 61 to 90 Days
		Opportunity Work	
		Schedule During Downtime or Shutdown	

Note: Priority is based on Equipment Criticality

FIGURE 5.5 Priorities and standard lead times.

WORK APPROVAL PROCESS

Maintenance notifications are reviewed for quality data to the accepted standard and to determine the proper risk-based priority and required end dates in a timely manner before moving to the work planning process. Ample time is given to perform work that is required to maintain the asset in an acceptable condition when the work will not result in unsafe conditions or the loss of revenue or will not exceed environmental regulations.

Ample time is given to plan, schedule, execute, and close the work order in time to meet the required end date. Proactive maintenance is achieved using the established priority and end date tools based on the risk of continuing to operate the equipment and completing the work before a projected failure date.

Urgent work is executed when there is immediate personal safety or environmental risk, risk of exceeding government regulation, or risk of the loss of production, breaking into the weekly schedule as part of the maintenance work management processes. Accommodation for this work is achieved by rescheduling the work assigned to the weekly schedule as determined by the asset owners.

In my experience, ownership and responsibility for the approval process are also with operations rather than maintenance because operations is responsible for the means of production, its criticality, and its priority to the business. Again, with the high level of operations involvement in notification or work order initiation, operations supervisors and managers coach the initiators to increase quality when it is below standard.

In smaller organizations, when staffing and structure do not permit, I have seen the work approval responsibilities delegated, but the delegates must be very involved with asset criticality and priority within the organization.

Process Design Critical Elements

The elements include:

- **NOTIFICATION QUALITY STANDARD.** As defined in the work identification process
- **SCREENING MEETING AGENDA.** A daily meeting to review and approve all notifications
- **RISK-BASED WORK SELECTION CRITERIA**
- **RISK MATRIX.** To identify safety, environmental, quality, and production risk
- **PRIORITY DEFINITIONS.** To determine the priority of all work in relation to all other work needed

- **PRIORITY LEAD TIMES.** The average acceptable standard length of time for work completion based on priority and type of work
- **RISK ASSESSMENT TOOLS.** A calculator used to determine the priority and required end date based on safety, environmental, quality, and production risk
- **PRIORITY SETTING CRITERIA/AUTHORITY.** Stakeholder(s) who can set, modify, and authorize work priority
- **SCHEDULE BREAK-IN CRITERIA.** Identify the quantifiable thresholds for authorizing schedule break-in work
- **SCHEDULE BREAK-IN RESPONSIBILITY.** Identifies who authorizes schedule break-in work that bypasses planning and scheduling
- **TARGET FINISH DATE CRITERIA.** Describes how target finish dates are determined
- **NOTIFICATION APPROVAL AUTHORITY.** Identifies the stakeholders who approve work to proceed

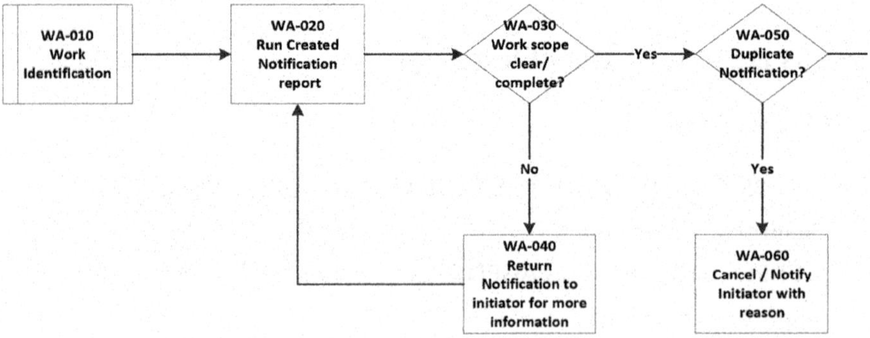

FIGURE 5.6 Work approval process flowchart.

Requirements/Expectations

All maintenance notifications are reviewed by qualified work screeners. These screeners review the notifications for quality data entry and required end date to the accepted standard and have the authority to reject the notification and return it to the initiator for coaching and correction to meet the standard.

It's critical that the proper risk-based priority and required end dates are determined by supervision or management and all notifications receive the proper authorized approval. All notifications are advanced in a timely manner to allow ample time for planning, scheduling, and execution to meet the required end date.

Figure 5.6 is a flowchart that traces the steps in the work approval process, and Figure 5.7 defines the steps.

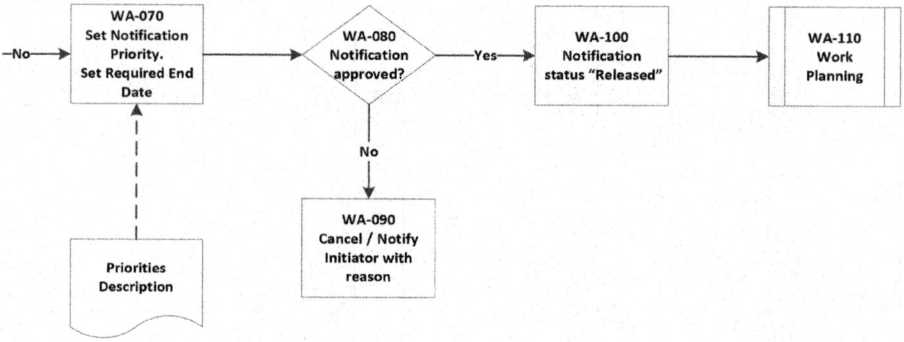

FIGURE 5.6 Work approval process flowchart (*continued*).

Process Title: **Work Approval**
Revision: **1.0**

Step Number	Step Description	Step Definition
WA-010	Work Identification	Notifications from the Work Identification process.
WA-020	Run Created Notifications report	Check in CMMS for all unapproved Notifications. Notifications queue should be checked every day.
WA-030	Work scope clear and complete?	Review Notifications for required fields
WA-040	Return Notifications to initiator for more information	If the Notification is not clear in scope, or is missing required information, return to initiator for more information.
WA-050	Duplicate Notifications?	Is the Notification a duplicate of another Notification?
WA-060	Cancel / Notify Initiator with reason	If there is an existing Notification, cancel the Notification and provide reason for rejection in the long text.
WA-070	Set Notification Priority. Set Required End Date	Based on the Priorities Description or other methods, assign the proper priority and Required End Date. Conduct Work Screening Meeting
WA-080	Notification approved?	Approve or reject W/O. Possible reasons for rejection: -Functionality -Cost -System Compatibility
WA-090	Cancel / Notify Initiator with reason	Enter reason for rejection into Notification long text. Set status to canceled and save Notification. The requestor will be automatically notified.
WA-100	Notification status "Released"	Set Notification to Released and send to the Planner's queue for processing.
WA-110	Work Planning	Released work orders go to the Work Planning process.

FIGURE 5.7 Definitions of steps in the work approval process.

The RACI chart in Figure 5.8 uses the information from Figures 5.6 and 5.7.

			Operations					Maintenance								RE	MRO		
RACI Chart for:	Work Approval	Operations Director	Superintendent	Production Area Lead	Production Area Coordinator	Gatekeeper	Shift Team Leader	Operator	Engineering and Maintenance Manager	Maintenance Manager	Maintenance Lead	Technician	Maintenance Planner	Maintenance Scheduler	Initiator	Reliability Engineer	Maintenance Buyer	Stockroom Clerk	Contractor
Step Number	Step Description																		
WA-010	Work Identification																		
WA-020	Run Created Notifications report			A		R							C						
WA-030	Work scope clear and complete?			A		R							C						
WA-040	Return Notifications to initiator for more information			A		R									R				
WA-050	Duplicate Notifications?			A		R													
WA-060	Cancel / Notify Initiator with reason			A		R									I				
WA-070	Set Notification Priority. Set Required End Date			A		R													
WA-080	Notification approved?			A		R													
WA-090	Cancel / Notify Initiator with reason			A		R									I				
WA-100	Notification status "Released"			A		R													
WA-110	Work Planning																		
	R = Responsible "The Doer"																		
	A = Accountable "The Buck Stops Here"																		
	C = Consult "Before a Decision"	Revision :	1.0																
	I = Inform "Keep in the Loop"																		

FIGURE 5.8 Work approval process RACI chart

Work Screening Process

It is preferable to conduct a work screening meeting on all unapproved notifications each day to determine the quality, priority, and required end date of the notification before approving and advancing it to the planning process. Each day that passes increases the likelihood of failure, resulting in a negative impact on safety and production. If you don't plan for correction of the incipient failure, the inevitable actual failure will schedule itself as a schedule break-in. It is preferred that less-than-quality notifications are returned to the initiator and that coaching is delivered to increase future quality. An example of a work screening meeting agenda is shown in Figure 5.9.

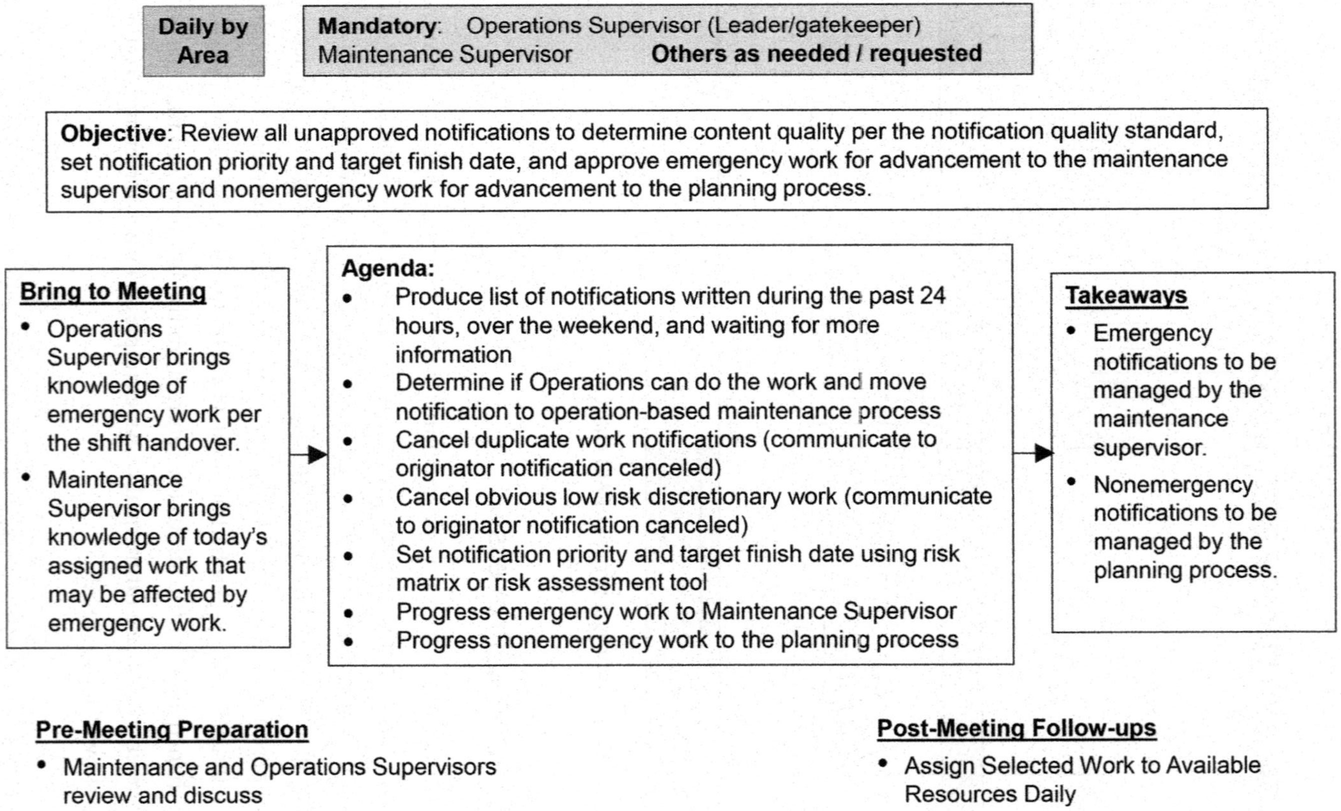

FIGURE 5.9 Work screening meeting agenda.

Risk-Based Priority

Risk-based prioritization is a method to rank possible failures by assessing their potential likelihood of occurring at a predictable time in the future and their impact if they do, focusing resources on high-risk items first to maximize equipment availability, minimize damage, and ensure timely action. It ranks work from high to low priority to provide a strategic approach to decide what needs attention and when.

Risk Matrix

A risk matrix is a visual tool that prioritizes potential risks of failure by plotting their likelihood (probability) against their impact on a grid, typically colored green for low priority, yellow for medium priority, and red for high priority to show which threats need urgent attention and to set required end dates for lower-priority work (see Figure 5.10). These required end dates are used during the work scheduling process to determine the optimized sequence of work across multiple work orders.

Some of the smaller companies I've worked with have a simple grid showing the priorities, the type of work to be completed, and a standard lead time to complete the work. They use the grid to set priorities and required end dates through the experience of the senior stakeholders. Just remember, scheduling is enabled by priorities and required end dates.

Probable time to breakdown (or equivalent)	Priority ratings					
Less than 2 days	7	3	2	1	1	1
2 to 7 days	7	4	3	2	1	1
7 to 30 days	7	5	4	3	2	1
30 days to 90 days	7	6	5	4	3	2
More than 90 days	7	6	6	5	4	3
Consequences: Health and safety	No effect	Safety Concerns only	First aid	Lost time injury	Multiple injuries	Fatality or permanent disability
Consequences: Environment	No effect	Environmental concern only	Internal Incident	Loss of Compliance (Dept down)	Loss of Compliance (Plant down)	Massive effect
Consequences: Total cost (production loss + maintenance cost)	No cost	<$10k	$10k - $100k	$100k - $1MM	$1MM - $10MM	> $10MM
Consequences: Customer, community or Investor relations	No effect	Concern only	Short term minor impact	Loss of Reputation with a few contacts	Loss of Reputation with many contacts	Business measuring reputation loss

	Completion targets						
Priority rating	7	6	5	4	3	2	1
No shutdown required	Consider rejecting	8 - 16 Weeks	4 - 8 Weeks	1 - 4 Weeks	Within 1 week	1 day	Immediate 24/7
Shutdown Required	Consider rejecting	When people available	Next 2 or 3 shifts	Next sched. shift	First opportunity	Start within 2 days	Immediate 24/7

FIGURE 5.10 Maintenance risk matrix.

Risk-Based Priority Calculation

Recently I had the opportunity to work on the implementation of a very well-known CMMS that has developed an integrated risk analysis tool by asking several questions about the work to be executed. It automatically entered the priority and required end date into the work notification. These could be modified based on the acceptable risk to the stakeholders.

At a company I consulted for, I worked with an engineer who developed a risk calculator that used a standard maintenance risk matrix combined with the company's standard priorities in a popular CMMS to calculate the priority and required end dates by answering five risk-based questions.

Target Finish Dates

When setting required end dates for work notifications or target finish dates for work orders, it's important to understand lead-time offsetting and target start dates for work orders. Each task has a duration and a sequence that work back in time from the target finish date to determine the target start date. Remember when setting the required end date or target finish date that there also needs to be time for planning, materials lead time, and scheduling. See Figure 5.11, which presents an example of target finish dates and lead-time offsetting.

As an added bonus, when a target finish date is exceeded, typically due to changing priorities, ask operations to perform another risk assessment on the work order and move the target finish to a future acceptable date or change the priority to a break-in if the risk of failure is high. Work cannot be scheduled and executed in the past.

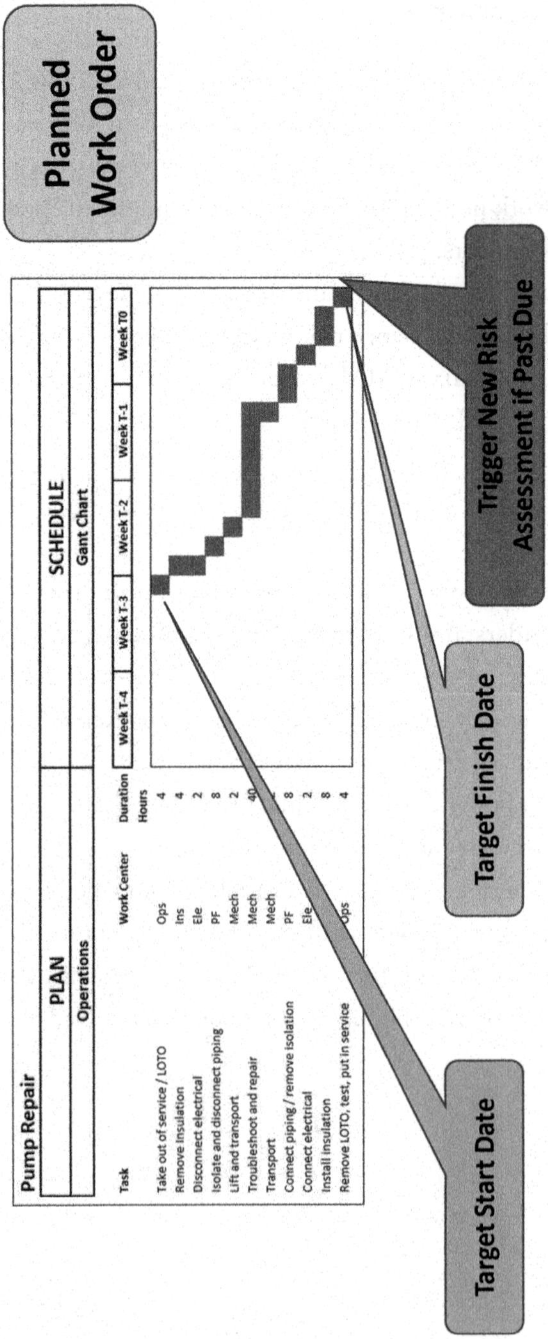

FIGURE 5.11 Target finish dates and lead-time offsetting.

WORK PLANNING PROCESS

Planning is the *what* and *how* for managing work execution. The *what* refers to the scope of the work to be accomplished, and the *how* is the sequence of tasks, materials, labor, information, and specialty tools needed to complete each piece of work.

A work plan is not a detailed instruction used to direct technicians on the completion of their work. It is a mechanism used to block out time on an integrated schedule so that expert technicians can do their jobs. If detailed instructions are needed, these can be added to the individual tasks' long descriptions. This can also be a detailed document attachment that is provided internally or by the asset manufacturer for a particular model of equipment.

Individual tasks in a work plan are designed to identify the task, the type of labor resource who will complete the task, how many of this particular labor resource are needed, and the duration of the task. A new task should be added to the work plan when a different labor resource discipline is needed to complete a task or when multiple labor resources are needed at the workface to work during the same time frame simultaneously or assisting each other. This also applies to heavy and specialty equipment with or without an operator or support craft who is needed to assist.

Appropriate work planning contributes to the overall safety during the execution of work. The planner will work to determine the requirements to complete the identified scope of work. The main components of the planning process include ensuring that the correct equipment, materials, people, and budgetary approvals are in place to proceed.

Typically, work planning is solely the responsibility of the maintenance planner, with the exception of break-in work going to the maintenance frontline supervisor. Planning independence is important as a check and balance with maintenance execution, to measure actual cost against planned cost for optimizing maintenance budgets. There are many examples where planners are physically separated from maintenance execution to isolate them from day-to-day schedule disruptions to ensure that a healthy backlog of planned work exists well into the future to optimize labor and materials cost.

I have worked in environments covering large geographical areas where it's not practical for planners to travel to every worksite to develop the scope of work and plan, such as with upstream well heads, compression and scrubbing sites, and pipelines in the oil and gas industry. In these cases, maintenance lead technicians were utilized for job scoping while executing work in the area where a plan was needed, with planners building the plan in the CMMS. Alternatively, I did work at one similar operation where planners scoped every job. The planners coordinated their need for scoping and travel routes to optimize their time and distance traveled, covering multiple sites per day.

Planners are often viewed as administrators, as planning can be a misunderstood process. Planning is a science that is reflected in the logic of well-established CMMSs. One of my greatest experiences was teaching a planning and scheduling class in a copper smelting operation where we agreed with the maintenance manager that we would train maintenance planners, maintenance schedulers, maintenance supervisors, and operations supervisors in a cross-functional class setting.

The supervisors were shocked at the details and difficulties experienced by the maintenance planners and schedulers. The biggest eye-opener was during a scheduling exercise where after the participants completed the manual time-consuming scheduling process of hard-to-schedule tasks, we disrupted the schedule with emotion-driven schedule break-in work and the supervisors had to choose which work to delay and adjust all the schedules. The classroom time limit ended before the rescheduling could be completed.

Understanding the Maintenance Planning Process and Its Importance

Maintenance planning is very much a misunderstood topic and underappreciated function. When completing my undergraduate degree in business, while working full-time for one of the largest carpenter's hand tools manufacturers in the world as a machine operator, I was asked what I wanted to do next. I replied that I wanted to start by being a production planner as they seemed like the spokes of a wheel that interacted with all facets of the plant. I was told, "No you don't. It's a thankless job, and you'll never get credit, only blame." I left that company and became a planner.

Planning is very detailed. Missing one tiny detail, like a small screw, can bring the whole operation to a standstill. Planning taught me how to put all the details together into one big finished picture, and how to take more than my fair share of criticism and lumps. It truly made me a stronger person who was ready to move toward manufacturing management. I've used my planning experience every day of my career, at and out of work.

Planning and Scheduling Constraints

While planning, the planner encounters many roadblocks. A specialized resource is not available, assets are not available due to production schedule changes, parts and materials are not available, tools and equipment are not available, budgeting and funding are not available, or contractors are not available when needed to perform their work. Someone who has never been a planner doesn't understand the details and underestimates the value of the planning process.

Maintenance Work Labor and Materials Estimating

Estimating provides an approximate judgment or opinion regarding the worth, amount, size, weight, etc., of all elements of maintenance work to calculate the expected costs as best as can be determined using the knowledge of the planner and other subject-matter experts when needed.

Estimating is used to predict cost prior to work execution, determine scheduling resources (CMMS and human capital), budget for needed cost controls, and measure actual hours worked versus estimated hours.

Estimates can be conceptual, relying on a thought or idea, dictated by a budget or using actual costs. Before the estimate can be utilized for scheduling, there must be a detailed plan for the work to be executed. Maintenance planning is a time-consuming process.

Estimating Techniques

These techniques include past experience, job classification/standardization, industrywide costs, labor standard, established job standards, and historical data.

Past Experience

The estimator determines the probable time required based on personal experience, knowledge, and ability. This type of estimate is easy to develop, the cost of estimating is low, and all jobs can be estimated. On the other hand, these estimates are inconsistent with variance between estimators, and training of estimators, making comparisons between estimators difficult.

Job Classification/Standardization

For estimating purposes, jobs can be classified and standardized. The classifying of jobs into a given type, cost, or time requirements creates consistency between estimators. This facilitates comparison to similar jobs and industry benchmarks.

Industrywide Costs

For estimating purposes, use maintenance cost per unit of product, per hour of operation, or per mile of mobile equipment as an average across an entire industry.

Labor Standard

Labor can be standardized as the amount of time required by an average qualified technician to perform a given task.

Established Job Standards

Agreed-upon standards can be used as guidelines to compare with similar jobs and are based on observations to determine actual work performed.

Historical Data

Historical data can be used for better consistency and lower cost after analysis of the execution of past work. It can be used by new planners and saves cost through repetition. However, using what has historically happened may be repeating past mistakes. When using historical data, alternative methods are difficult to compare and may not be a true reflection of new or future work.

In my many years as a planner, I have learned that no estimating technique or planning style works in all situations. Typically, all these techniques are used to compliment each other as applied by the individual planner.

When evaluating and training planners as a group, I stick to the science of planning and recognize that planning *is* a science and keeping it simple is the best approach. I work with all the planners collectively to develop a standard work plan format and standard writing techniques that produce a schedule that is uniform in look and function.

Process Design Critical Elements

Elements include:

- **NOTIFICATION REJECTION AUTHORITY.** Planners must have the authority to reject notifications or work orders for poor quality.
- **PLANNING BACKLOG PRIORITY MANAGEMENT.** Work prioritization is an integral part of backlog management. Each responsible stakeholder has to manage the sequence of their work to meet work notification and work order priority, required end dates, and target finish dates.
- **JOB COMPLEXITY DEFINITION/CRITERIA.** To provide guidelines for maintenance planners to develop adequate job plans for a particular type of job (see Figure 5.12).

  Examples:

  - **SIMPLE.** Typically a single craft task that should last less than one shift has minimal planning requirements to effectively execute and involves only minimal permitting. Field preparation activities are typically handled by the maintenance execution team.

- **BASIC.** An intermediate complexity task that requires more than one craft to execute and requires planning support and typically operations preparation.
- **COMPLEX.** A task that has a high degree of execution difficulty and integrates multiple crafts, technical support, support planning input, process preparation, and sequencing of tasks.

- **WORK ORDER QUALITY STANDARD.** Typically using the work notification quality standard.
- **WORK ORDER PLANNING STANDARD.** A detailed list of elements needed for job complexity levels, e.g., drawings, safety standards, specific equipment, and contractor support.
- **PLANNING CHECKLIST.** For basic and complex jobs, reflecting the work order planning standard elements and used for taking field notes.
- **COST APPROVAL THRESHOLDS.** Tailored to each organization, shows authorized stakeholders who approve maintenance work plans for advancement to supply chain, scheduling, and execution based on a planned cost roll-up for the work order.

	Requirements		Plan Components
SIMPLE	• Minimal SHE (can be handled by the Maintenance Supervisor on the day of the job) • No hot work/ confined space/entry permit req'd. • Clear work scope without job site visit • Materials are either free issue materials or common warehouse items (planner must be certain warehouse will have them in stock) • Work requires one shift or less • Only one craft is required • Job is done frequently by the technician and supervisor • No major equipment is needed (cranes, etc.)	PLANNING	• 1 CMMS task • Planner estimates job (costs and hours) and assigns resource (gets PO if required) • Planner approves job and makes RTS • No "work package". Supervisor gets his info from W/O only
MEDIUM	Everything else	MORE PLANNING	• Field visit required • Planning checklist required • Task out the job clearly (including sequencing): • 1 craft per task, 1 job step per task • Tasks named clearly • May require quotes / bids • "Package" required • Safety precautions highlighted • Check for bundling opportunities • Documentation: - Bill of Materials - Pictures - MSDS - As found / as left documentation Forms for specific procedures - Eng Documentation - Marked up drawings - LOTOTO List - Waste disposal forms
COMPLEX	• Above normal SHE considerations: - Formal JSA required - Electrical live work permit required - Formal risk assessment - Variance to safety procedure required • Process planning is complex: - Formal process plan req'd - Rate cut or shutdown needed - More than 1 shift of Operations prep • Permitting is complex • Multiple planners required for the job • 1st time execution (never been done before) • Specialized equipment needed (>65-ton crane, etc.) • Complex execution	LOTS OF PLANNING	ALL JOBS: • Process tasks stepped out in SAP • Detailed task sequencing • Written job plan / scope / special procedures • Specialized PPE requirements • Cost / Estimate optimization (lump sums, estimating standards, bundling of work, etc.) IF APPLICABLE: • Formal estimating tools • Equipment, History documentation • Specialized equipment details • Lift Plan • Risk assessment • Loading diagrams • MOC

NOTES:

1) Simple Jobs MUST meet ALL the requirements. Any one requirement that's not met bumps it up to a Medium Job
2) If any requirement for a COMPLEX job is met, it must follow a COMPLEX plan. If it does not meet any of them, it's a Basic Job
3) Plan components are additive (Basic jobs require everything that simple jobs do. Complex jobs required all the components of a Basic plan)

FIGURE 5.12 Planning job complexity guidelines.

Maintenance Planning Requirements

In their normal course of business, maintenance planners verify the quality of the notifications in their backlog; prioritize their backlogs based on priorities, required end dates, and known long lead-time materials and labor constraints; and complete a site visit to determine the scope of the work needed and create the work order from the work notification.

My personal coaching to planners in the field has been to complete a planning checklist while examining the asset and recording as many details as possible while visualizing the work being executed (see Figure 5.13). This is also a good time to inspect the immediate area for safety considerations and other assets needing attention, keeping in mind that it's also their responsibility to enter notifications for other incipient failures observed.

While observing the asset, identify the main components of the plan, including required tasks, task sequence, required labor, and the duration required for each task. If needed, where materials, parts, and components identification allow, record the materials and components needed for each task. This information will be valuable when planning the job in the office setting, ensuring no time is lost returning to the asset. This effort reduces the time needed in front of the CMMS.

In addition, in the CMMS, identify in the plan the asset-specific work instructions and documentation, specialty equipment needed, scheduling constraints, and safety measures needed for each task. After completion of the plan, obtain budgetary approvals to proceed.

LOCATION/UNIT________________ **WORK ORDER** _________________ **DATE**_____________
WORK ORDER DESCRIPTION____

Work Package Complexity: Basic [ ] Complex [ ]

Safety / Health / Environmental/PPE
- [ ] Caustic/acid
- [ ] Heat protection
- [ ] Breathing air
- [ ] NORM (Natural Occurring Radioactive Material)
- [ ] Benzene
- [ ] Asbestos
- [ ] Lead
- [ ] Proximity to high voltage lines
- [ ] Material Safety Data Sheet (MSDS
- [ ] Job Safety Analysis
- [ ] Standard Practices for Wastes
- [ ] Scrap Material Management
- [ ] Planning for Proper Handling of Excavated Material
- [ ] Waste Handling (Disposal and Transfer)
- [ ] Evaluate Waste for Processing In-House
- [ ] Container Labeling
- [ ] Confined Space
- [ ] Underground obstruction/interference
- [ ] Special PPE ____________________
- [ ] Other________________________

2) Work Area and general
- [ ] Walking surface
- [ ] Work areas clean and ready to work
- [ ] Special cleanup activities planned
- [ ] Repair tag located
- [ ] Other related repair tags in the area for bundling?
- [ ] Process equipment shutdown/start-up needs
- [ ] Limited space access
- [ ] Over side/over water

3) Scaffolding
- [ ] Scaffolding required
- [ ] Removal planned
- [ ] Height above deck/height above ground ____________
- [ ] Special needs (enclosed for welding, fire hazards, etc.)

4) Insulation
- [ ] Checked for asbestos (exchanger/pipe/sight-glass gaskets)
- [ ] Remove Insulation
- [ ] Asbestos removal (notification and environmental requirements)
- [ ] Reinstallation of insulation planned
- [ ] Heat Tracing

5) Painting
- [ ] Check for lead paint
- [ ] Lead abatement planned
- [ ] Painting required (Special Coating)

6) Procedures
- [ ] Work procedures (site specific)
- [ ] Safety Standard
- [ ] Electrical Procedures
- [ ] Blind list attached
- [ ] Hot Work required
- [ ] MOC attached
- [ ] Welding procedures
- [ ] Equipment preparation procedure
- [ ] Lifting Plan
- [ ] MIQA
- [ ] Stress Relieving / PWHT
- [ ] Temporary hoses?
- [ ] Neutralization
- [ ] Variances
- [ ] Road Closure

7) Materials, Equipment, and Drawings
- [ ] Bill of materials complete
- [ ] POs complete
- [ ] Large or rental equipment vendors notified
- [ ] Warehouse stock material available
- [ ] P&IDs and other drawings required
- [ ] Plot plans required
- [ ] Photos, sketches available
- [ ] Line specs available
- [ ] Equipment drawings available
- [ ] Vendor Documentation

8) Inspection and Testing
- [ ] Positive material inspection required
- [ ] WQCF
- [ ] Special cleaning requirements needed (blasting, brushing, etc.)
- [ ] Radiography required
- [ ] Hydrotest required (Water, Soda Ash wash, etc.)
- [ ] Electrical Testing

9) List Specialty Contractors: (e.g. leak repair, hydro blasting, etc.)

____
____
____
____

10) Equipment needed:
- [ ] Crane (size)____________________
- [ ] Welding Machine
- [ ] Air Compressor
- [ ] Portable Light Plant
- [ ] Breathing Air Bottles
- [ ] Manlift / Scissor lift / porta-scaffold\DROT
- [ ] Extractor or other maintenance tool planned
- [ ] Rope Access
- [ ] Torqueing Equipment
- [ ] Misc.:____________________.

11) Support Craft needed: (e.g. I&E, Machinist, Rigging, etc.)
- [ ] E&I support (e.g. disconnect / reconnect)
- [ ] Machinist (e.g. uncouple motors)
- [ ] Field mechanics
- [ ] Laborers

12) Other Documents
- [ ] Equipment history available
- [ ] Work Order History
- [ ] Project Schedules (e.g. P3, Microsoft, etc.)
- [ ] Job Cost Estimate
- [ ] Other historical references applicable ______________

13) Estimating Considerations
- [ ] Lump sum bidding
- [ ] Budgetary Quotes
- [ ] Historical Costs
- [ ] BRRF Estimating tool
- [ ] Industry Estimating standards (e.g. Page of Nations)
- [ ] Bundle with other work

FIGURE 5.13 Maintenance planning checklist.

Figure 5.14 is a flowchart that traces the steps in the work planning process, and Figure 5.15 defines the steps.

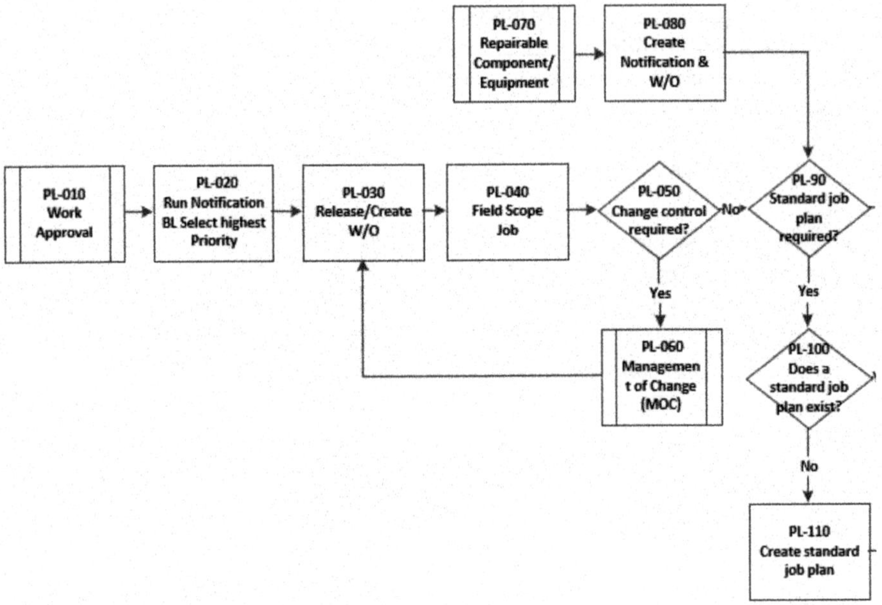

FIGURE 5.14 Work planning process flowchart.

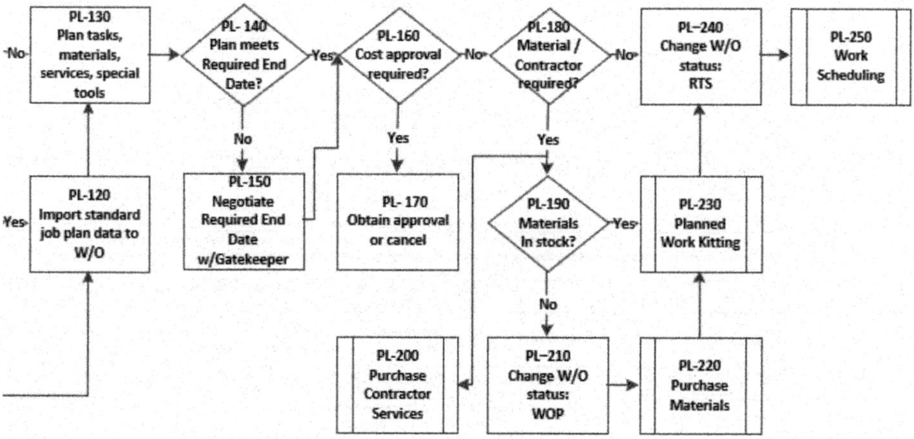

FIGURE 5.14 Work planning process flowchart (*continued*).

Process Title: **Work Planning**
Revision: **1.0**

Step Number	Step Description	Step Definition
PL-010	Work Approval	Approved notifications from Work Approval process.
PL-020	Run Notification BL Select highest Priority	Check Notification backlog. Released notifications are assigned to proper Planner group. Work orders are prioritized based on notification Priority and Required End Date.
PL-030	Release/Create W/O	Review scope/quality of notification and create work order.
PL-040	Field Scope Job	Visit the jobsite, consult with initiator and appropriate SMEs as necessary. The planning checklist should capture all required information.
PL-050	Change control required?	Assess if any aspect of the job requires change control, per Change Control process.
PL-060	Management of Change (MOC)	Refer to MOC process.
PL-070	Repairable Component/Equipment	Refer to Repairable Component/Equipment process.
PL-080	Create Notification & W/O	Create notification & work order for Repairable Component/Equipment.
PL-090	Standard job plan required?	Criteria for Job Plan (at discretion of Planner): significant planning effort, repetition of work, and complexity of job.
PL-100	Does a standard job plan exist?	Is there an existing standard job plan in the library?
PL-110	Create standard job plan	Create standard job plan to asset as identified on original work order. Attach to all similar assets for future use.
PL-120	Import standard job plan data to W/O	Attach all aspects of the standard job plan to the work order. This should include operations, parts/materials, equipment, labor, contractors/services, etc.
PL-130	Plan tasks, materials, services, special tools	Plan all: tasks, contractors, services, work centers, labor hours, drawings, commissioning/startup requirements, sequence of operations. Update standard job plan data as needed.
PL-140	Plan meets Required End Date?	Based on parts/materials delivery and contractors/services availability, determine if work order can execute before the Required End Date.
PL-150	Negotiate Required End Date w/Gatekeeper	If parts/materials delivery or contractors/services availability cannot meet execution before the Required End Date, renegotiate with the Releasing Gatekeeper.
PL-160	Cost approval required?	Seek approval above the Planner's approval limit.
PL-170	Obtain approval or cancel	If approval is not granted, request cancellation of the work order.
PL-180	Material / Contractor required?	Does this work order require parts/materials, services/contractors, specialty PPE, or tool crib tools in order to complete?
PL-190	Materials in stock?	If materials are in stock, pick list will be released to the stockroom.
PL-200	Purchase Contractor Services	If services/contractors are required, progress requisition to Purchasing.
PL-210	Change W/O status: WOP	Update status of work order to WOP (Waiting on Parts).
PL-220	Purchase Materials	If materials are not in stock, progress requisition for direct purchase to work order.
PL-230	Planned Work Kitting	Materials will be kitted per the work order pick list the week prior to execution.
PL-240	Change W/O status: RTS	Update status of work order to RTS (Ready to Schedule).
PL-250	Work Scheduling	Refer to Scheduling process.

FIGURE 5.15 Definitions of steps in the work approval process.

The RACI chart in Figure 5.16 uses the information from Figures 5.14 and 5.15.

		Operations						Maintenance								RE	MRO		
RACI Chart for:	**Work Planning**	Operations Director	Superintendent	Production Area Lead	Production Area Coordinator	Gatekeeper	Shift Team Leader	Operator	Engineering and Maintenance Manager	Maintenance Manager	Maintenance Lead	Technician	Maintenance Planner	Maintenance Scheduler	Initiator	Reliability Engineer	Maintenance Buyer	Stockroom Clerk	Contractor
Step Number	**Step Description**																		
PL-010	Work Approval																		
PL-020	Run Notification BL Select highest Priority									A			R						
PL-030	Release/Create W/O									A			R		C				
PL-040	Field Scope Job									A	C		R		C				
PL-050	Change control required?									A			R						
PL-060	Management of Change (MOC)																		
PL-070	Repairable Component/Equipment									A			R						
PL-080	Create Notification & W/O									A			R						
PL-090	Standard job plan required?									A			R						
PL-100	Does a standard job plan exist?									A			R						
PL-110	Create standard job plan									A	C		R						
PL-120	Import standard job plan data to W/O									A			R						
PL-130	Plan tasks, materials, services, special tools									A	C		R						
PL-140	Plan meets Required End Date?					I				A			R						
PL-150	Negotiate Required End Date w/Gatekeeper					C				A	C		R				C		C
PL-160	Cost approval required?	C	C	C		C			C	A/C	C		R						
PL-170	Obtain approval or cancel									A			R		I				
PL-180	Material / Contractor required?									A			R				C		C
PL-190	Materials in stock?									A			R					C	
PL-200	Purchase Contractor Services									A							R		
PL-210	Change W/O status: WOP									A			R	R				C	
PL-220	Purchase Materials									A							R		
PL-230	Planned Work Kitting									A								R	
PL-240	Change W/O status: RTS									A			R	R				C	
PL-250	Work Scheduling																		

R = Responsible "The Doer"
A = Accountable "The Buck Stops Here"
C = Consult "Before a Decision"
I = Inform "Keep in the Loop"

Revision : **1.0**

FIGURE 5.16 Work planning process RACI chart.

WORK SCHEDULING PROCESS

As planning is the *what* and *how*, scheduling is the *who* and *when*. During scheduling, individuals or groups of like labor resources are assigned to each task and placed in a specific time frame on the calendar, typically by day in a weekly schedule to meet the target finish date while utilizing the labor resources fully. It is understood that individual tasks in a job plan may not be executed one after the other as time passes, that the job may sit idle while waiting for a particular labor resource to become available to complete the next task.

Maintenance Schedule Cost Reduction

The main focus of maintenance planning and scheduling is cost optimization. Scheduling ensures that assets are available when maintenance arrives to begin work, that the material and parts are provided where and when needed, that travel time is reduced between jobs, that there is better coordination for specialty technicians, and that outside services are only used when necessary.

An optimized maintenance schedule minimizes job interruption by reducing false starts, reducing or eliminating stops due to material and/or services availability, and ensuring that only a limited few labor resources are pulled from scheduled work to mitigate emergencies.

Maintenance and operations maintain continuity, ensuring asset availability is coordinated and asset downtime and start-up issues are reduced due to maintenance.

Scheduling Tools

Maintenance scheduling to the task level is a complex function that increases with the size of the organization. In medium-to-large organizations, a CMMS is necessary to manage the high volume of tasks needed to complete all the work.

To enable the scheduling function, an ample supply of planned, ready-to-schedule work orders is needed to fill the backlog. The work order needs a defined clear scope, all engineering needs to be complete, skills must be assigned, estimated cost and duration are determined, material/parts are available, required safety items are prepared, and outside services are coordinated and available.

In a controlled and managed work environment, activities from various stakeholder groups are aligned in consolidated schedules to achieve improved safety performance while maximizing production and resource effectiveness. Through the integrated planning and scheduling processes, activity requirements are continually refined as areas progress through a 90-day integrated scheduling process to a 30-day integrated scheduling process and to a ready-for-execution, seven-day scheduling process. Workforce availability, materials availability, and job priority are all considered when creating the schedules.

As the owner of the assets, the operations team typically has the final say in setting the maintenance schedule. The operations team works in conjunction with maintenance, as maintenance scheduling administers the process by pulling all scheduling data, placing all work due on the schedules by the predetermined target finish dates, and negotiating the final schedules between maintenance and operations.

The operations team commits to access to the assets, whereas maintenance commits to providing labor, information, materials and parts, and equipment needed to perform the work. During the weekly scheduling meeting, operations and maintenance enter into an agreement for maintenance activities to be completed the following week. It's important to keep in mind that scheduling and execution are not finite. A schedule is only valid the second it is completed. There are many variables that will come into play as the schedule is executed through time. If you have 100% schedule compliance, question if there is enough work on the schedule or if the planned tasks contain excessive duration. Unexpected scheduled break-in work will occur in most cases.

Process Design Critical Elements

Elements include:

- **AVAILABLE LABOR CALENDAR.** Is typically provided by human resources or the maintenance supervisor to identify who will be onsite during the execution scheduled time frame

- **WEEKLY SCHEDULING PROCESS CALENDAR.** Shows the scheduling process by day expressed by morning and afternoon activities
- **WEEKLY SCHEDULING MEETING AGENDA.** Reflects activities and responsibilities for the scheduling process stakeholders
- **MULTI-MONTH (THREE-MONTH) OUTLOOK MEETING AGENDA.** Expands on the weekly scheduling meeting agenda to add other stakeholders and a longer scheduling horizon, e.g., supply chain, engineering, and projects.

Weekly Scheduling Process

The scheduling process, if all plans are high quality, is a task that can be greatly simplified using well-known scheduling software that interfaces with planning CMMS software. These software tools, using the target finish dates and duration of the work order tasks, place the tasks on the calendar, taking into consideration the available labor needed for execution after checking the available labor calendar. Keep in mind that complete work orders may not be scheduled sequentially through time, that work on a particular work order may start and stop based on the availability of various labor, and that tools or special equipment is needed for the individual tasks. Also, work plan tasks can be locked for a particular time frame depending on when the needed resource or resources, due to other commitments, are available. The other tasks will be scheduled to accommodate this resource constraint.

The scheduler reviews the planned, ready-to-schedule corrective, preventive, and predictive work orders created by the system and creates a schedule (list) of the work orders available within a certain time frame. This list is sequenced by the target finish date, asset criticality, highest priority, and greatest estimated hours. Consider that large jobs may flow over to multiple weeks.

The scheduler continues until all available resources have been scheduled. The first-pass schedule is reviewed and approved by the operations and maintenance people who negotiate a week 1 schedule acceptable to all stakeholders. When an organization becomes proficient with a week 1 schedule, it expands, adding a week 2 preliminary schedule. The acceptable week 1 schedule is frozen using a snapshot in time and is used to measure work execution and schedule performance.

In my experience, the weekly scheduling process for the next week (week 1) is achieved in four days in the current week. On the morning of day 1, scheduling pulls all data for the seven-day period following the current seven-day schedule by target finish date and priority and balances the labor for the upcoming week 1 schedule for each day of the schedule. This labor balancing may indicate that too many or not enough labor resources exist and provides this finding to operations and maintenance for negotiations.

On the afternoon of day 1 and the morning of day 2, operations and maintenance review the information for approval or adjustment. On the afternoon of day 2, operations and maintenance collaborate and agree on their proposed week 1 schedule.

As the current week passes, adjustments are made to the week 1 schedule based on the performance to this week's schedule, and this information is provided to scheduling to make adjustments. This continues until the afternoon of day 3 when scheduling meets with operations and maintenance to agree on the week 1 schedule.

Adjustments continue until midday on day 4 if needed. On day 4 the schedule is "locked," which is typically a snapshot used to compare actual maintenance execution with the week 1 schedule, providing key performance indicator (KPI) data for continual improvement. The week 1 schedule is distributed to all stakeholders.

Once organizations are comfortable with a week 1 schedule, they typically expand to a week 2 prospective schedule updated weekly. This schedule utilizes the work with target finish dates for the week following the time frame of the week 1 schedule. It is known that the week 2 schedule carries a higher risk of not being realized, as priorities change in time and material and parts available in-house are typically not considered until it rolls to a week 1 schedule. It does create the need to consider the coordination of incoming parts and materials to be available when the week 2 schedule becomes the week 1 schedule. Once the week 2 schedule is widely used, organizations expand scheduling to a 30-day to 90-day large job and long lead-time materials job schedule.

Figure 5.17 is a flowchart that traces the steps in the work scheduling process, and Figure 5.18 defines the steps.

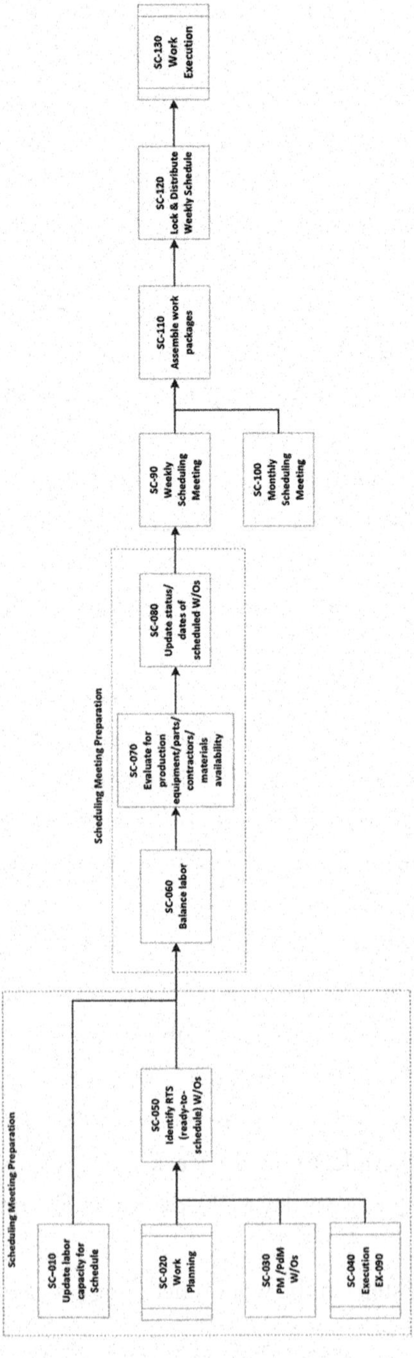

FIGURE 5.17 Work scheduling process flowchart.

Process Title: **Work Scheduling**
Revision: **1.0**

Step Number	Step Description	Step Definition
SC-010	Update labor capacity for Schedule	Labor capacities determined for each trade available for this schedule.
SC-020	Work Planning	Refer to Planning process. Typically covers CM (corrective) and rebuild work orders.
SC-030	PM /PdM W/Os	Include in this schedule Preventive Maintenance work orders e.g., PM/PdM.
SC-040	Execution EX-090	Refer to Execution process. See Reschedules and follow-up work orders.
SC-050	Identify RTS (ready-to-schedule) W/Os	Identify work orders that are "Ready to be Scheduled" and sort by priority.
SC-060	Balance labor	Add "ready to schedule" work orders to PM work orders to align labor requirements with labor availability.
SC-070	Evaluate for production equipment/parts/contractors/materials availability	Before adding work orders, ensure equipment availability. Ensure all parts/materials/contractors are onsite 1 week before inclusion on the schedule.
SC-080	Update status/dates of scheduled W/Os	Schedule Start/Finish dates are applied to WOs within the schedule.
SC-090	Weekly Scheduling Meeting	Once Schedule is balanced for labor and verified for equipment availability with the Maintenance Lead and Production Area Coordinators, it is confirmed.
SC-100	Monthly Scheduling Meeting	Conduct monthly meeting to review/evaluate all long-range, long-lead time, low-priority work.
SC-110	Assemble work packages	Work order packages are assembled according to Maintenance Lead. Publish all scheduled packages weekly.
SC-120	Lock & Distribute Weekly Schedule	At the end of the weekly scheduling cycle, lock & distribute the schedule and work packages.
SC-130	Work Execution	Refer to Work Execution process.

FIGURE 5.18 Definitions of steps in the work scheduling process.

The RACI chart in Figure 5.19 uses the information from Figures 5.17 and 5.18.

		Operations							Maintenance							RE	MRO		
RACI Chart for:	Work Scheduling	Operations Director	Superintendent	Production Area Lead	Production Area Coordinator	Gatekeeper	Shift Team Leader	Operator	Engineering and Maintenance Manager	Maintenance Manager	Maintenance Lead	Technician	Maintenance Planner	Maintenance Scheduler	Initiator	Reliability Engineer	Maintenance Buyer	Stockroom Clerk	Contractor
Step Number	**Step Description**																		
SC-010	Update labor capacity for Schedule									A	C			R					
SC-020	Work Planning																		
SC-030	PM /PdM W/Os									A			C	R					
SC-040	Execution EX-090																		
SC-050	Identify RTS (ready-to-schedule) W/Os									A				R					
SC-060	Balance labor									A				R					
SC-070	Evaluate for production equipment/parts/contractors/materials availability			A	R					A	R			C			C	C	C
SC-080	Update status/dates of scheduled W/Os									A				R					
SC-090	Weekly Scheduling Meeting			A	R					A	R			C			C		C
SC-100	Monthly Scheduling Meeting			A	R					A	R			C			C		C
SC-110	Assemble work packages									A	I			R					
SC-120	Lock & Distribute Weekly Schedule		I	I	I	I	I		I	A/I	I			R			I	I	I
SC-130	Work Execution																		
	R = Responsible "The Doer" A = Accountable "The Buck Stops Here" C = Consult "Before a Decision" I = Inform "Keep in the Loop"	Revision : 1.0																	

FIGURE 5.19 Work scheduling process RACI chart.

Weekly Scheduling Process Calendar

		Monday	Tuesday	Wednesday	Thursday
MX Supervisor	AM	Provide 2 week's Resources Pool minus training, off-site, vacation, etc. Notify Scheduler of WO's not completed last week	Review Two Week Schedule with Ops Supervisor		
	PM		Attend Preliminary Scheduling Meeting Bring inputs from schedule review	Provide updates since Preliminary Scheduling Meeting at Weekly Scheduling Meeting	Review work packages for Monday, walk down and assign jobs.
Scheduler	AM	Review 2 week's Resources Pool minus training, off-site, vacation, etc.		Review impact of changes on Weekly Schedule. Based off of Preliminary Scheduling Meeting	
	PM	Reset info on jobs not complete Generate Two Week Schedule	**Preliminary Scheduling Meeting:** Review Two Week Schedule. Note and make changes to schedule accordingly	Facilitate **Weekly Scheduling Meeting:** Record adjustments from Ops and MX Supervisors	**Lock and publish 2 Week Schedule**
Ops Supervisor	AM		Review Two Week Schedule with MX Suoervisor		
	PM		Attend Preliminary Scheduling Meeting Bring inputs from schedule review	Provide updates since Preliminary Scheduling Meeting at Weekly Scheduling Meeting	

FIGURE 5.20 Weekly scheduling process calendar (days of the week may vary).

Weekly Scheduling Meeting Prerequisites

Prior to the weekly scheduling meeting, the maintenance frontline supervisor reviews and understands the scope of the work to be completed on all work orders on the week 1 schedule and verifies that the labor, materials, parts, and tools needed to complete the work are available onsite. This typically includes contractor labor and services and their access to the facility.

The supply chain team is typically not personally involved in the weekly scheduling meeting, as all materials and components are verified onsite before work is placed on the week 1 schedule. Working with a week 2 schedule affords time for the maintenance and supply chain teams to track needed materials and components two weeks in advance.

To develop week 1 through 90-day schedules, there needs to be an ample amount of planned work. In my experience, the greatest economies are realized with a six-week backlog of planned work in addition to the larger, long lead-time jobs. This highlights the need to separate the planners from the day-to-day execution of scheduled and break-in executable work to focus on maintaining a planned backlog of work that keeps all labor resources busy for their total onsite hours for each week.

An often-overlooked benefit is the six weeks of parts, components, and materials in the pipeline that can be pulled forward if priorities change, especially in cases where a schedule break-in occurs. In this case, it's likely that big cost savings are achieved, avoiding the need for expediting and thus incurring expensive accelerated shipping costs. In many cases, I've been present when the needed parts or materials are already onsite.

An example of a maintenance weekly schedule is shown in Figure 5.21. An example of a weekly scheduling meeting agenda is shown in Figure 5.22.

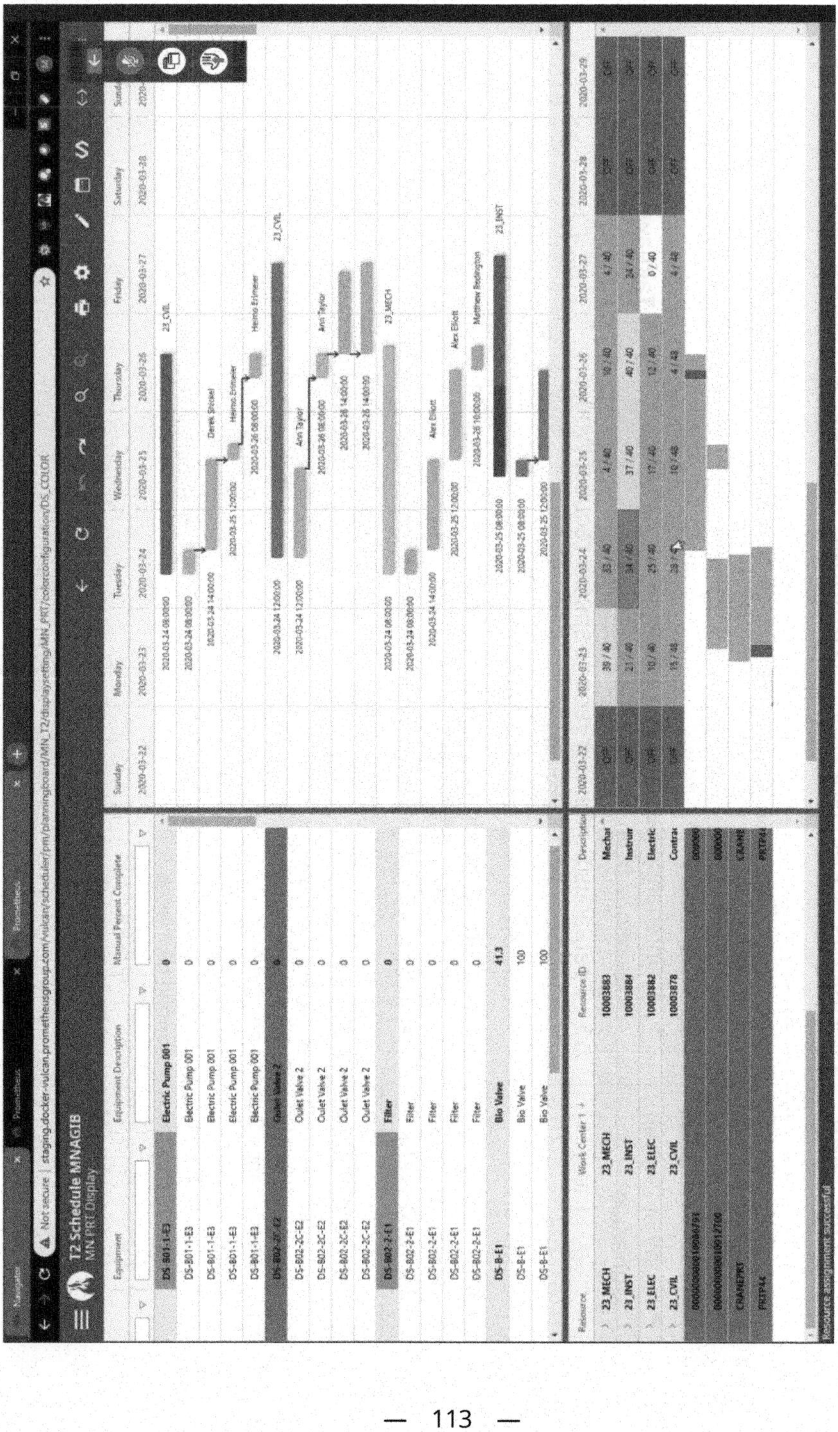

FIGURE 5.21 Maintenance weekly schedule (Prometheus® Screenshot).

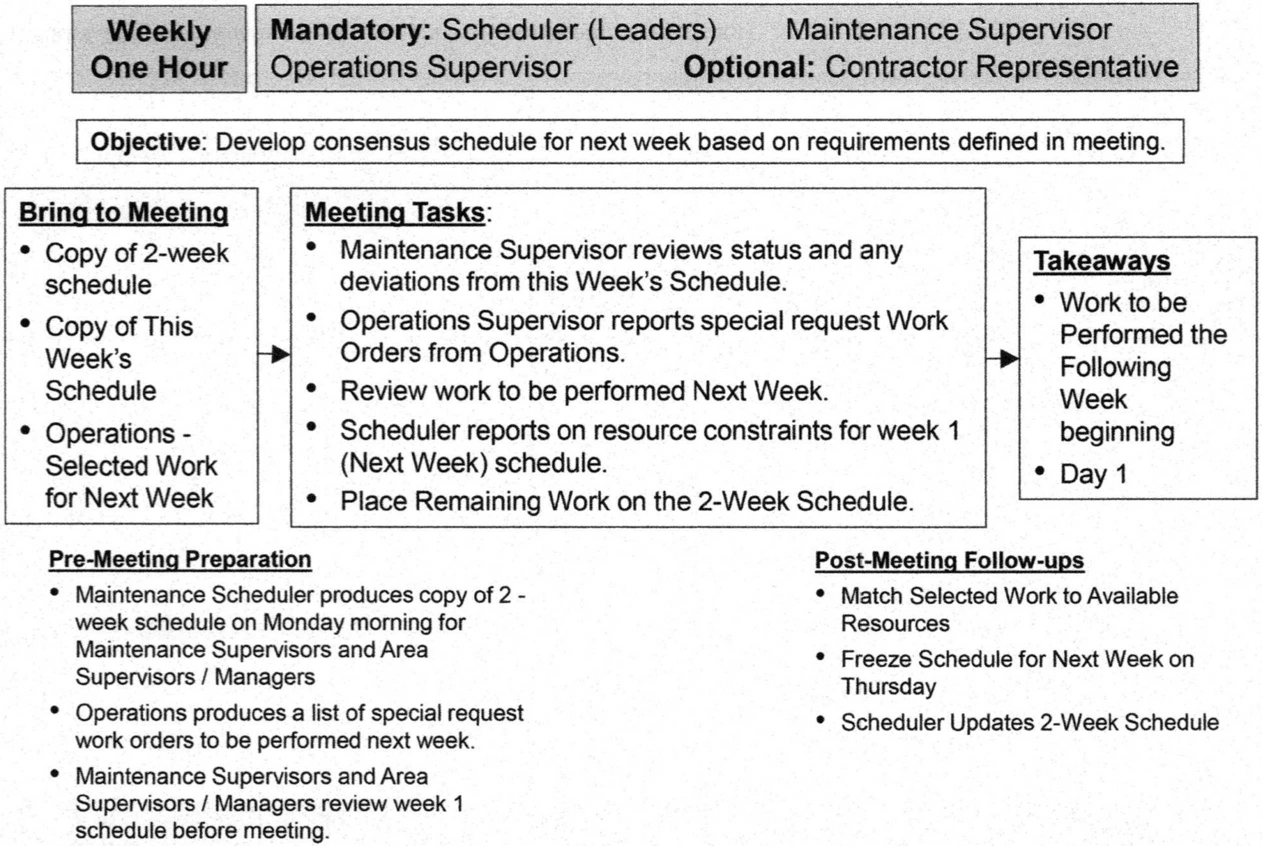

FIGURE 5.22 Weekly scheduling meeting agenda.

Multi-Month (3-Month) Outlook/Scheduling Meeting

The multi-month scheduling meeting process feeds the weekly scheduling meeting process. This meeting examines the larger jobs and jobs affected by long lead-time materials and components and typically includes the participation of project management, engineering, and supply chain personnel.

Figure 5.23 shows a multi-month outlook/scheduling meeting agenda.

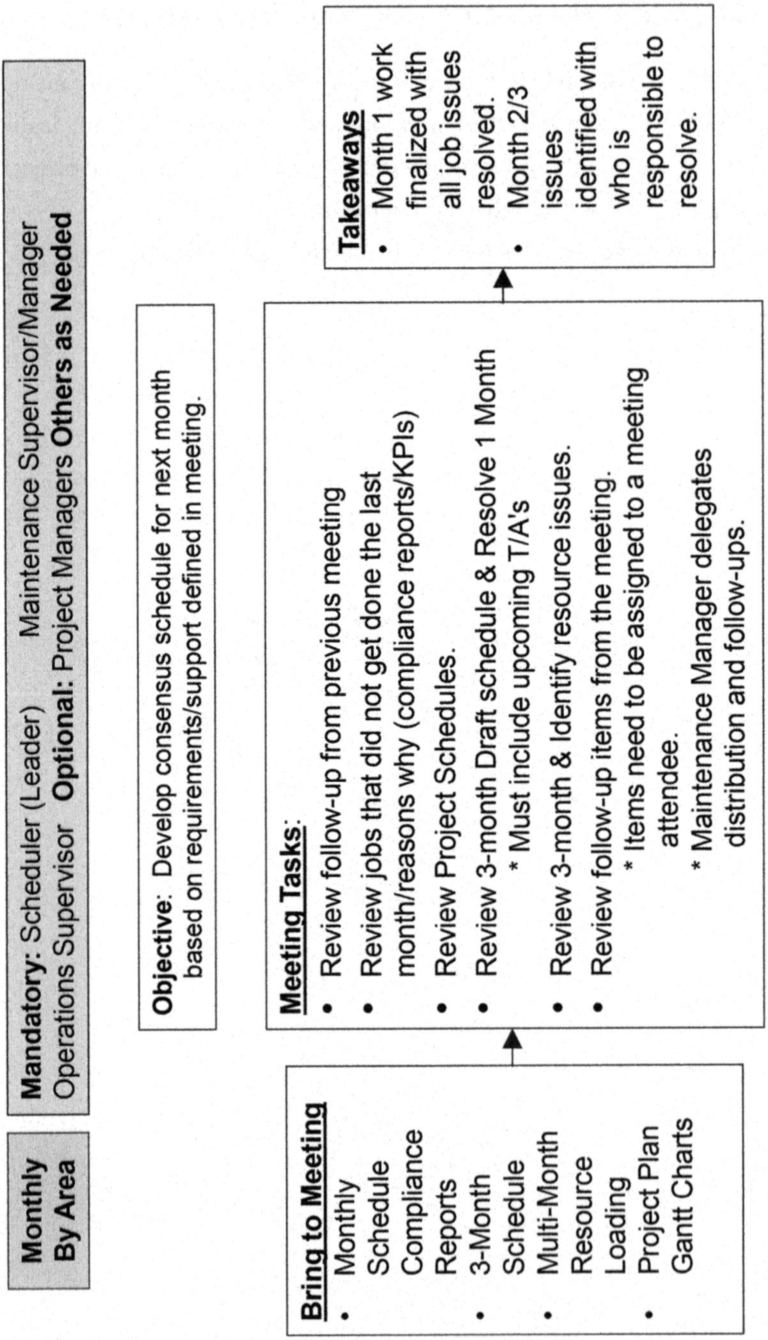

FIGURE 5.23 Multi-month (3-month) outlook/scheduling meeting agenda.

WORK EXECUTION PROCESS

Work execution is where the tire meets the road. All activities before work execution are performed to facilitate the smooth execution of work each day. All previous activities are designed to ensure that labor resources are utilized to the full extent.

Work is executed on a day-to-day basis as previously planned and scheduled on the approved week 1, 7-day schedule. Work packages are provided to ensure understanding of the work to be accomplished, including all permits, drawings, and information as specified on the work order. All required materials, tools, and equipment are planned for and provided in advance of work commencement.

If urgent "break-in" work is required that was not on the schedule, then previously scheduled work will need to be deferred or else additional workers utilized to complete the additional work. Typically, the decision to break work into the week 1 schedule comes from and is authorized by operations, effectively changing the priorities of scheduled work. As the owners of the assets, the operations team is responsible for choosing to add extra labor or defer work and selects the work to be deferred. In all cases, work must be completed safely and according to regulatory and organizational requirements.

As schedule break-in work bypasses the planning, supply chain, and scheduling functions, the materials and labor cost still need to be tracked in the work order for historical data and analysis. Typically the maintenance frontline supervisor is responsible for the planning and execution of the schedule break-in work. As brief as the execution of break-in work is, it is still planned in an abbreviated process with the maintenance frontline supervisor scoping the work through observation and determining the needed resources to complete the work. At a minimum, the maintenance frontline supervisor enters and issues materials and labor to the plan in the CMMS.

Typically, maintenance is the primary owner of work execution; however, as owners of the assets, the members of the operations team have the authority to change the schedule, which is then reflected in the schedule compliance KPIs for cause and improvement. The operations team provides direction for the rescheduling of work displaced when its priorities create a schedule change. This authority also extends to the daily assignment of work.

Process Design Critical Elements

Elements include:

- **DAILY SCHEDULING/DAILY WORK ASSIGNMENT MEETING AND AGENDA.** The daily meeting where operations and maintenance meet to discuss the work in progress and agree on the work to be executed over the next 24 hours.
- **DAILY WORK ASSIGNMENT TOOL/CMMS CRITERIA.** Specifies the use of a manual work assignment tool or the use of the CMMS functionality if available.
- **DAILY WORK ASSIGNMENT RESPONSIBILITY.** As the owner of the work execution process and resources, the maintenance frontline supervisor typically performs the daily assignment of work to their direct reports.

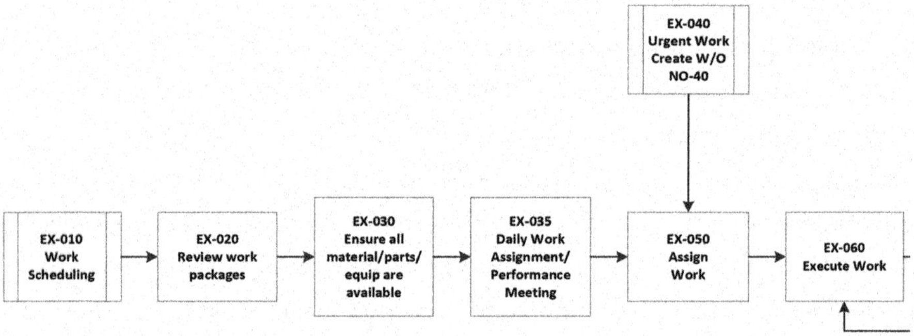

FIGURE 5.24 Work execution process flowchart.

- **WORK ORDER COMPLETION CONFIRMATION STANDARD.** After the completion of work, the maintenance technician typically records the actual time, verifies the damage/failure and cause codes, and briefly describes the work completed for reliability analysis and completes the required reports.
- **BARRIER DATA COLLECTION STANDARD.** Using a manual or CMMS work assignment process, the maintenance supervisor typically records the barriers encountered during work execution for analysis and resolution.

Figure 5.24 is a flowchart that traces the steps in the work execution process, and Figure 5.25 defines the steps.

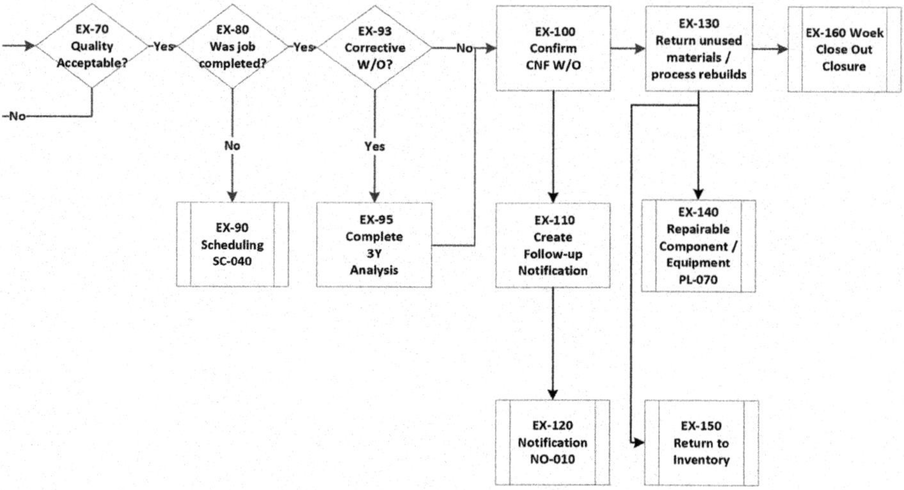

FIGURE 5.24 Work execution process flowchart (*continued*).

Process Title: **Work Execution**
Revision: **1.0**

Step Number	Step Description	Step Definition
EX-010	Work Scheduling	Refer to Work Scheduling process.
EX-020	Review work packages	Scheduled work packages for the week and next day are reviewed.
EX-030	Ensure all material/parts/equip are available	Make provisions to have materials/parts delivered to the worksite if needed.
EX-035	Daily Work Assignment/Performance Meeting	This meeting reviews the last 24 and next 24-hour schedules and performance to the last 24-hours' work assignment.
EX-040	Urgent Work Create W/O	Refer to Notification process. Urgent work receives a minimum of planning including labor and parts and is coordinated and documented by the Maintenance Lead.
EX-050	Assign Work	Mechanics assigned to work orders for completion.
EX-060	Execute Work	Perform work as per tasks on W/O. On completion of work: -Remove LOTOTO -Verify operation
EX-070	Quality Acceptable?	Review for operating condition and quality of work. Back to EX-060 Execute work for rework performed until quality and equipment operation are acceptable.
EX-80	Was job completed?	Were assigned WO tasks completed?
EX-090	Scheduling	For work order tasks not completed, determine future week for completion and return to the scheduling process. Return unused material to staging area to be held until job is rescheduled and continued.
EX-093	Corrective W/O?	Was this work executed to correct a failure?
EX-095	Complete 3Y Analysis	Determine the root cause of the failure using the 3Y (Why) method questioning the nature of the failure. Enter the root cause in the header Long Description of the work order.
EX-100	Confirm CNF W/O	Enter hours worked. Enter detailed work completion information in work order logs (planning/engineering feedback). Identify PM/PdM Job Plan changes. Close out permits.
EX-110	Create Follow-up Notification	Complete follow-up Notification for continuation of this work if needed or for unrelated failures observed in the vicinity of this work.
EX-120	Notification	Follow-up Notifications are channeled to the beginning of the process.
EX-130	Return unused materials / process rebuilds	For completed jobs, issue materials used on work order if not yet issued. Tag unused materials for return to inventory. Return tool crib items. Tag rebuildable components for processing.
EX-140	Repairable Component / Equipment	Deliver rebuildable components / equipment for processing.
EX-150	Return to Inventory	Deliver unused materials to inventory.
EX-160	Work Close Out	Refer to Work Close Out process.

FIGURE 5.25 Definitions of steps in the work execution process.

The RACI chart in Figure 5.26 uses the information from Figures 5.24 and 5.25.

		Operations							Maintenance							RE	MRO		
RACI Chart for:	Work Execution	Operations Director	Superintendent	Production Area Lead	Production Area Coordinator	Gatekeeper	Shift Team Leader	Operator	Engineering and Maintenance Manager	Maintenance Manager	Maintenance Lead	Technician	Maintenance Planner	Maintenance Scheduler	Initiator	Reliability Engineer	Maintenance Buyer	Stockroom Clerk	Contractor
Step Number	**Step Description**																		
EX-010	Work Scheduling																		
EX-020	Review work packages									A	R								
EX-030	Ensure all material/parts/equip are available									A	R							C	
EX-035	Daily Work Assignment/Performance Meeting									A	R			R					
EX-040	Urgent Work Create W/O									A	R								
EX-050	Assign Work									A	R	C/I							
EX-060	Execute work					I					A	R							R
EX-070	Quality Acceptable?					C					A	R							R
EX-80	Was job completed?					C				A	R	C							S
EX-90	Scheduling																		
EX-93	Corrective W/O?										A	R							
EX-95	Complete 3Y Analysis										A	R							
EX-100	Confirm CNF W/O										A	R							
EX-110	Create Follow-up Notification					C					A	R							
EX-120	Notification																		
EX-130	Return unused materials / process rebuilds										A	R						C	
EX-140	Repairable Component / Equipment										A	R							
EX-150	Return to Inventory										A	R							
EX-160	Work Close Out																		
	R = Responsible "The Doer"																		
	A = Accountable "The Buck Stops Here"																		
	C = Consult "Before a Decision"	Revision :		1.0															
	I = Inform "Keep in the Loop"																		

FIGURE 5.26 Work execution process RACI chart.

Daily Scheduling/Daily Work Assignment

Typically, the week 1 schedule is broken down into daily work buckets to meet the expected labor available for each day. One common practice is to hold a daily meeting to discuss maintenance execution performance for yesterday and today and to agree upon work to be executed through tomorrow.

Once in place with operations and maintenance understanding their roles, these meetings occur in a very short time frame and can include a brief analysis of the barriers to work completion encountered, as most operations and maintenance organizations communicate in real time throughout the day as situations develop and need resolving. See Figure 5.27.

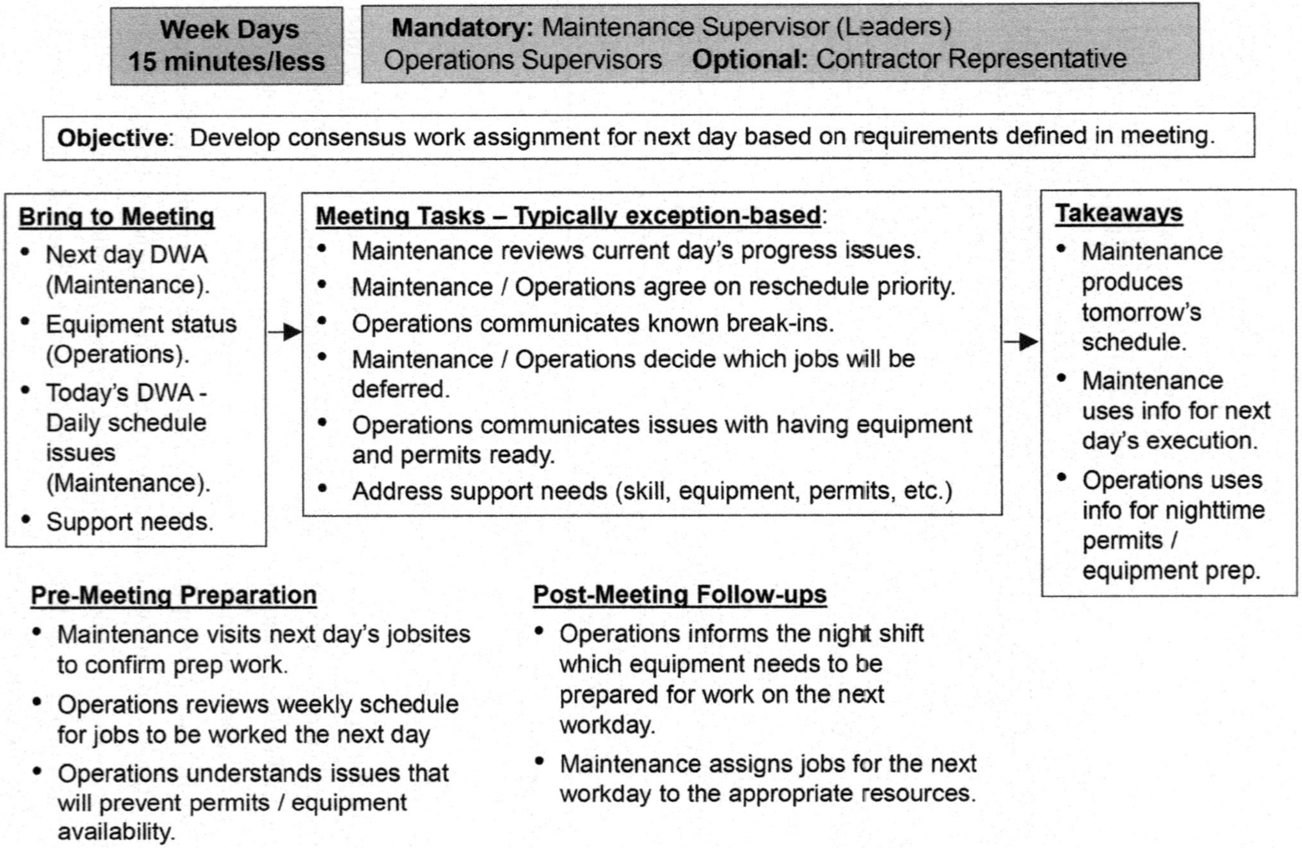

FIGURE 5.27 Daily scheduling/daily work assignment meeting agenda.

The current state of maintenance scheduling software, in some cases, automates the daily work assignment process, making it very easy for a maintenance supervisor to assign work to the individual labor resources by name and collect daily barrier data to achieve the work at the work order level. Tracking this information is also efficient using common, simple manual tools. The data collected can improve processes and assets and optimize the utilization of labor, achieving very high maintenance cost savings that most supervisors and managers might not expect.

An example of a CMMS daily schedule is shown in Figure 5.28.

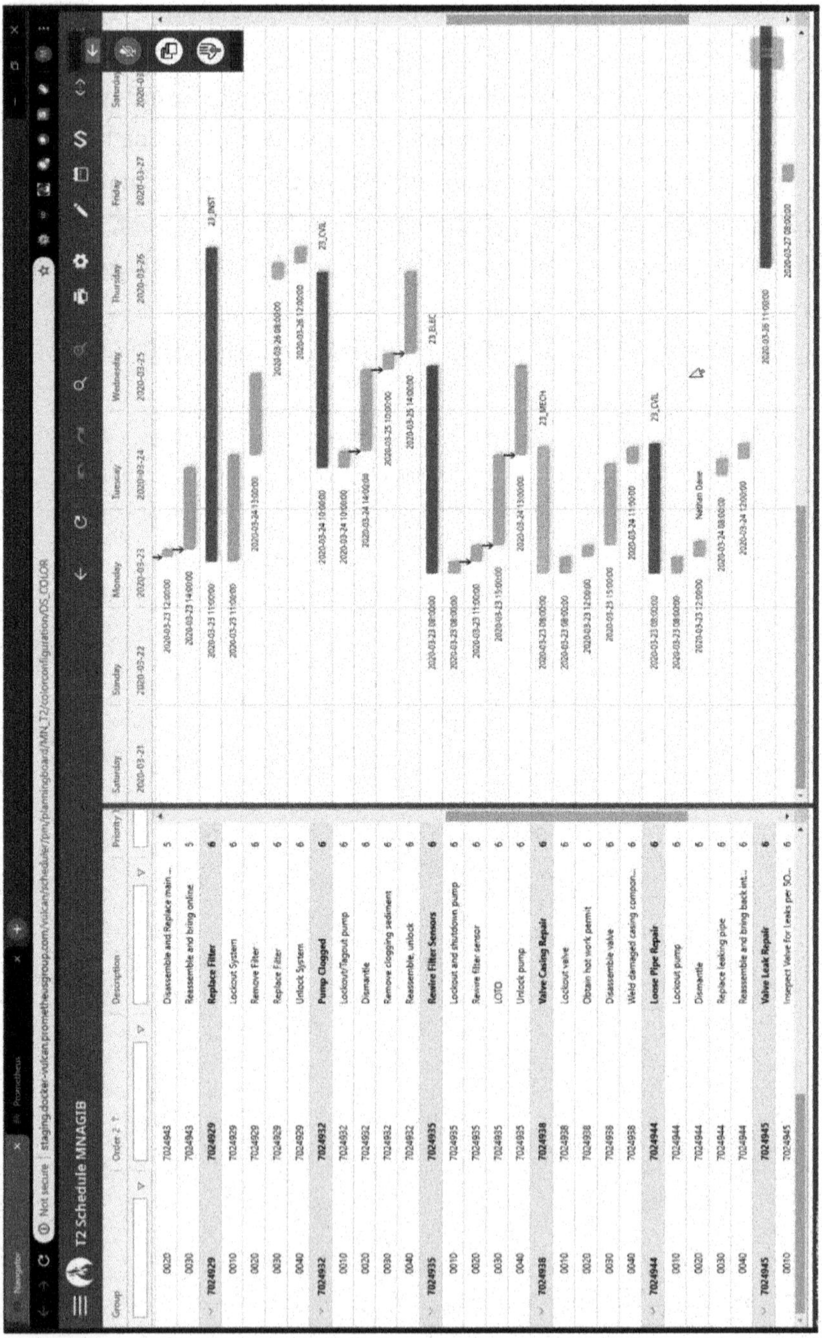

FIGURE 5.28 CMMS daily schedule. (Prometheus® Screenshot)

Work Execution Expectations

Plans are never perfect. When plans are incorrect or insufficient, the most efficient, cost-effective, and timely course of action is to proceed with the work, with leadership and decision-making by a maintenance frontline supervisor adding needed materials and labor to complete the job and work order.

At the conclusion of the work execution, unused material and parts are returned to stock for use in other jobs, barriers are documented, operations accepts the work, actual hours worked and materials are recorded in the work order, and any feedback for planning and engineering is communicated.

Potential improvements to the job package are provided to planners and engineers through notes, damage and cause codes, and reasons for any deviations. Repairable items are tagged and prepared for repair, follow-up items identified and documented, and schedules updated.

Daily Scheduling/Daily Work Assignment Tool

When CMMS functionality doesn't exist, a simple daily work assignment tool can be created using a spreadsheet that resembles the departmental dispatch boards that I used early in my career to make the schedule visible to all team members for scheduling machine tools by displaying them in a central location. Basically, the sheet has work orders and task descriptions down the left side with a list of labor resource names across the top. Planned hours and actual hours are tracked in the boxes corresponding to the work orders assigned to the labor resources, taking care to assign enough work to fill 100% of the resources' day.

Actual hours recorded in the corresponding boxes are used to calculate variances. Variances by hours are tracked to the right to a standard set of barriers, and simple statistics are calculated at the bottom. The barriers are prioritized by frequency for targeted resolution and elimination. See the example in Figure 5.29.

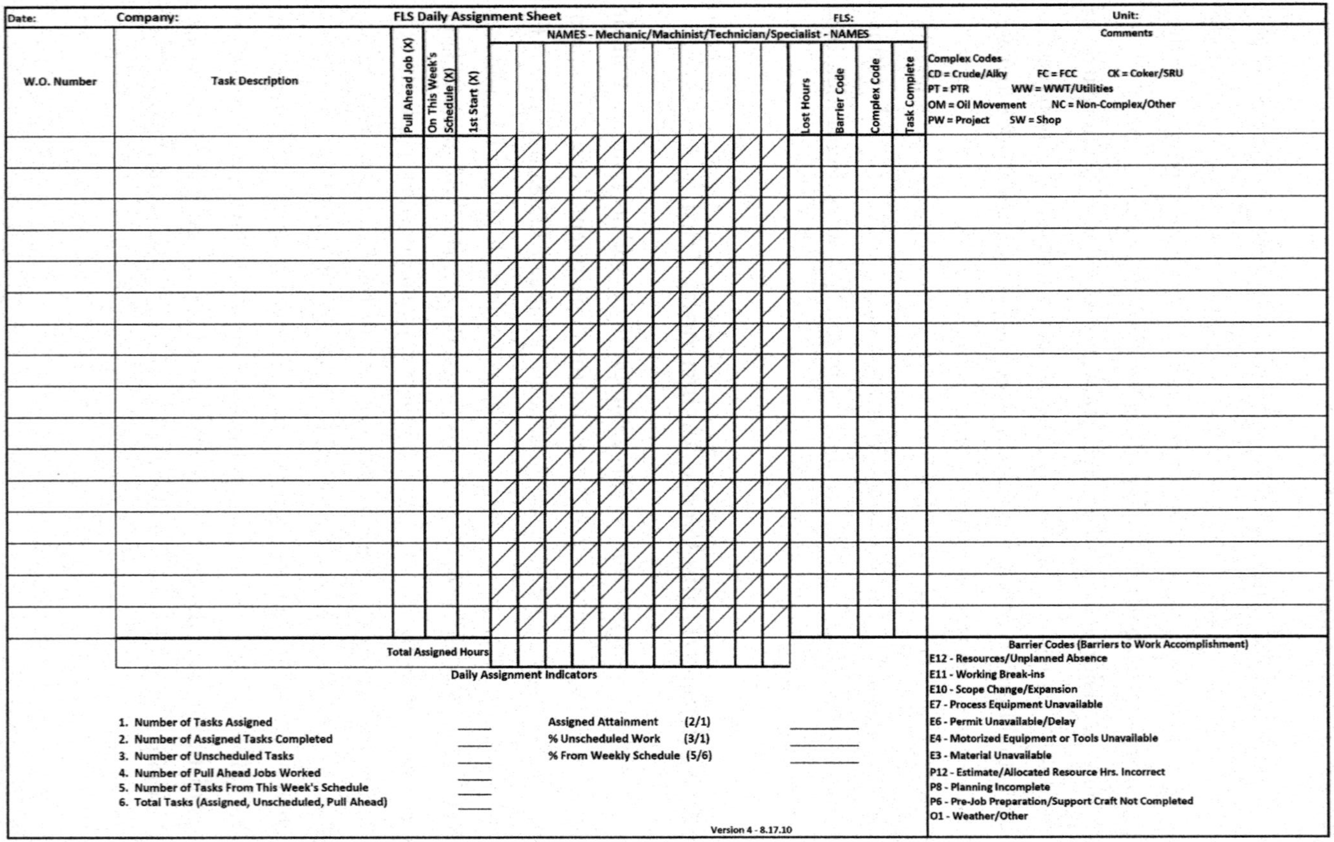

Date: | Company: | FLS Daily Assignment Sheet | FLS: | Unit:

W.O. Number	Task Description	Pull Ahead Job (X)	On This Week's Schedule (X)	1st Start (X)	NAMES - Mechanic/Machinist/Technician/Specialist - NAMES	Lost Hours	Barrier Code	Complex Code	Task Complete	Comments

Complex Codes
CD = Crude/Alky FC = FCC CK = Coker/SRU
PT = PTR WW = WWT/Utilities
OM = Oil Movement NC = Non-Complex/Other
PW = Project SW = Shop

Total Assigned Hours

Daily Assignment Indicators

1. Number of Tasks Assigned ____
2. Number of Assigned Tasks Completed ____
3. Number of Unscheduled Tasks ____
4. Number of Pull Ahead Jobs Worked ____
5. Number of Tasks From This Week's Schedule ____
6. Total Tasks (Assigned, Unscheduled, Pull Ahead) ____

Assigned Attainment (2/1) ____
% Unscheduled Work (3/1) ____
% From Weekly Schedule (5/6) ____

Barrier Codes (Barriers to Work Accomplishment)
E12 - Resources/Unplanned Absence
E11 - Working Break-ins
E10 - Scope Change/Expansion
E7 - Process Equipment Unavailable
E6 - Permit Unavailable/Delay
E4 - Motorized Equipment or Tools Unavailable
E3 - Material Unavailable
P12 - Estimate/Allocated Resource Hrs. Incorrect
P8 - Planning Incomplete
P6 - Pre-Job Preparation/Support Craft Not Completed
O1 - Weather/Other

Version 4 - 8.17.10

FIGURE 5.29 Daily work assignment tool.

Work Execution Barriers

During the normal course of work execution, priorities change due to unforeseen events like equipment breakdowns that are critical to continued production operations. The occurrence of these events and their causes are a major source of data for continual improvement. Again, some CMMSs track this data at the work order level.

This data is tracked daily for work that is not completed to the daily assignment of work. Eliminating the easily identifiable causes and preventing reoccurrences happen in real time on a shorter interval than the weekly schedule. The daily roll-up of common occurrence data in a weekly format is investigated for root cause, measured in lost hours and categorized to a predetermined standard, and reported on a weekly basis during a short weekly barrier team meeting.

Common barriers experienced include unplanned absence, break-in job, scope change, estimating error, equipment not ready, parts not available, tool/machine not available, permitting delay, support craft not available, operations delay, production priority change, etc., based on the relevance to the organization.

Daily Work Assignment Tool Data Collection and Usage

As a normal course of action, during the day-to-day execution of maintenance work, barriers to completing the day's work are encountered, highlighted, discussed, troubleshot, and resolved. This is fixing problems as opposed to the long-term fixing of assets and processes. As discussed, many CMMSs have the capability to track these barriers and produce statistics for continual improvement for high-frequency reoccurrences.

Not every organization has a CMMS that can track and report these statistics. With the daily work assignment tool, the daily performance at the bottom of the sheet can be easily consolidated daily at a weekly level and can be used to drive a weekly barrier team meeting to discuss, troubleshoot, and resolve recurring, larger systemic barriers (see Figure 5.30).

This meeting typically uses the same stakeholders as the daily scheduling/daily work assignment meeting and can follow that meeting once per week at the beginning of the week for the previous week's barriers. Collect data and identify trends for use in the work execution barrier team meeting (see Figure 5.31).

DWA Weekly / Monthly Summary

January 2020

Prepared Date:: 2/19/2025

Area: ABC Company

Supervisor: John Doe

DWA SUMMARY

Indicators	W/E 1/10/20		W/E 1/17/20		W/E 1/24/20		W/E 1/31/20		W/E 2/7/20		TOTAL
Number Assigned Tasks Completed	49		67		47		44				207
Number Tasks Assigned	56		90		66		53				265
Number Schedule Breaks	23		33		37		21				114
% Assigned Attainment	88%		74%		71%		83%		#DIV/0!		78%
% of Schedule Breaks	47%		49%		79%		48%		#DIV/0!		55%
Barrier Hours	HRS	#	HRS	#	HRS	#	HRS	#	HRS	#	Total HRS
A - Unplanned Absence			2	2	9.5	5	1.5	1			13
B - Break-In Job	5	3	35	16	28.5	13	15.5	10			84
C - Scope Change	1.5	1									2
D - Estimating Error	4	3	1.5	3	0.5	1	8.5	9			15
E - Equipment Not Ready											0
F - Parts Not Available					5	1					5
G - Tool/Machine Not Available											0
H - Permitting Delay											0
I - Support Craft Not Available			1	1							1
J - Operations Delay											0
K - Production Priority Change											0
TOTALS	10.5	7	39.5	22	43.5	20	25.5	20	0	0	119

Top 2 Barriers

Comments:

Based on 21 Days

2,068 Annual Lost Hours

Equals 1 Person

FIGURE 5.30 Work execution barrier tracking.

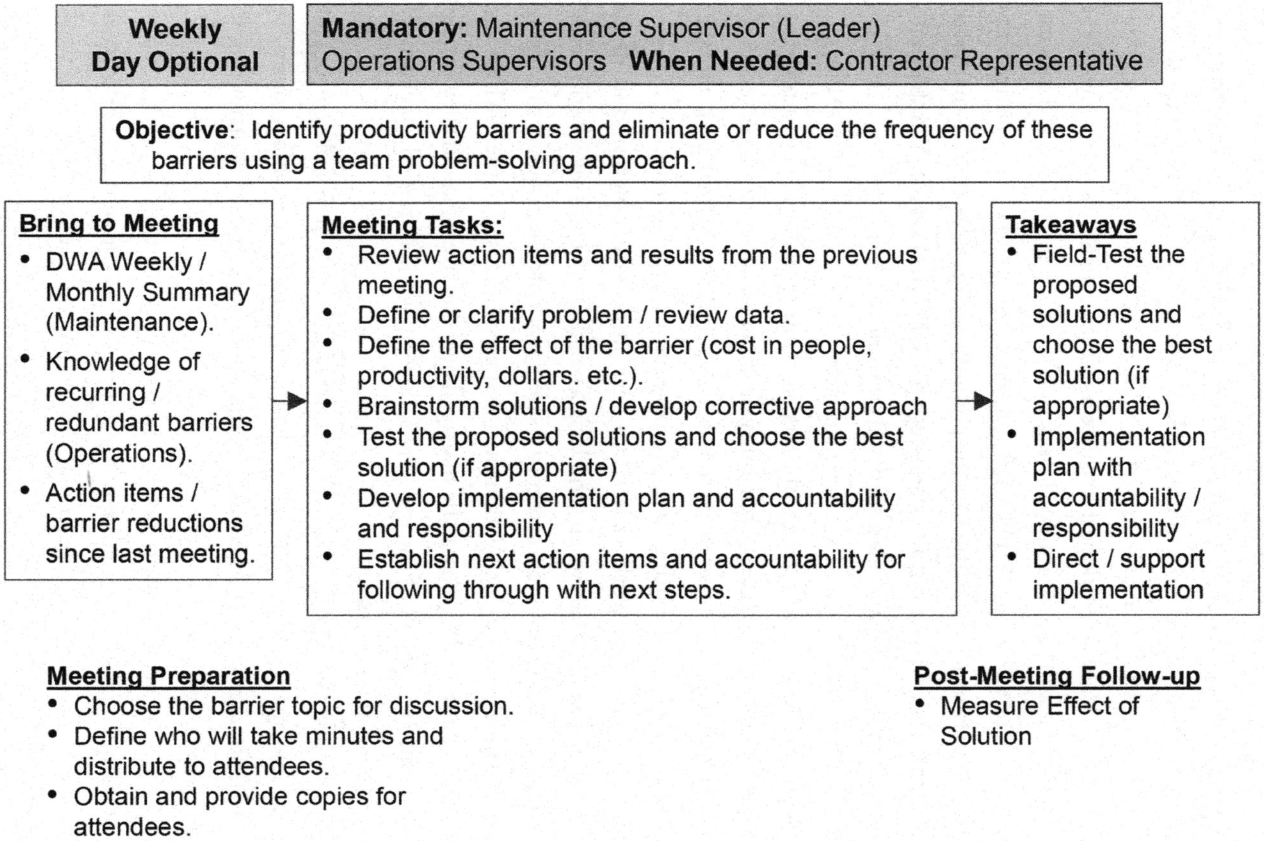

FIGURE 5.31 Work execution barrier team meeting agenda.

WORK CLOSE OUT PROCESS

Once the work execution is completed, the work order completion feedback to the maintenance supervisor is reviewed, and any follow-up work is used to create new work notifications. All pertinent work details and reports are filed, and the work order is closed. Job details are now available for analysis for future improvement and optimization. Work performance and maintenance work management KPIs are tracked and used to improve work performance and the MWM process.

Data is required before technical completion of work orders pertaining to the actual work performed, condition of the equipment, hours worked, actual work activities completed, planning and engineering feedback, equipment strategy and job plan upgrades, and materials issued and consumed.

This information is used to increase equipment reliability, develop and improve reliability strategies, control costs, and improve the MWM processes. It's important to note the value of reliability data analysis utilizing the object part/equipment type, failure/damage type, and cause for improving the health and reliability of the assets contributing to maximize equipment availability.

The completion of work orders is a very important function and often neglected. I first noticed this when scrubbing a backlog to identify work still needing completion, identifying two category 1 safety work orders in danger of causing imminent death in a refinery. Work orders that are not closed in the CMMS are subject to churn and added cost to be managed over and over again. If they are out of the system, they are out of the way, and follow-up work, if needed, is identified and planned in a new work order.

Next to safety on the list of priorities is the lack of collection and reporting of reliability information. I've seen many reliability departments unable to optimize equipment strategies due to the lack of failure and work history contributing to excessive asset downtime.

Job planning and reliability details are now available for analysis for future optimization. Generally the maintenance supervisor, as the steward of the work execution process is responsible for data collection and the completion of work orders.

Process Design Critical Elements

Elements include:

- **WORK ORDER COMPLETION STANDARD.** Data collection is defined and standardized, and responsibility for collection is assigned.
- **WORK ORDER COMPLETION REVIEW CRITERIA.** Maintenance frontline supervisor data sampling for coaching and continual improvement.

Figure 5.32 is a flowchart that traces the steps in the work close out process, and Figure 5.33 defines the steps.

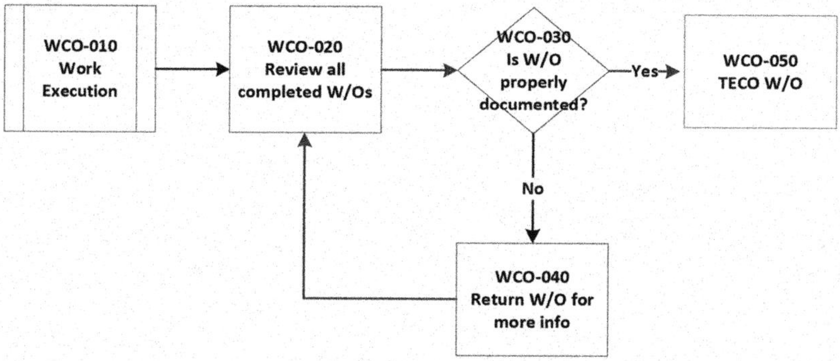

FIGURE 5.32 Work close out process flowchart.

Process Title: **Work Close Out**
Revision: **1.0**

Step Number	Step Description	Step Definition
WCO-010	Work Execution	Refer to Work Execution process.
WCO-020	Review all completed W/Os	Review all work orders with status of Confirmed.
WCO-030	Is the W/O properly documented?	Hours worked. Detailed work completion information in work order logs (planning/engineering feedback). Identified PM/PdM Job Plan changes. Parts / materials issued.
WCO-040	Return W/O for more info.	Work order is returned to mechanic to complete correctly.
WCO-050	TECO W/O	TECO Technically Complete work order in CMMS.

FIGURE 5.33 Definitions of steps in the work close out process.

The RACI chart in Figure 5.34 uses the information from Figures 5.32 and 5.33.

		Operations							Maintenance							RE	MRO		
RACI Chart for:	Work Close Out	Operations Director	Superintendent	Production Area Lead	Production Area Coordinator	Gatekeeper	Shift Team Leader	Operator	Engineering and Maintenance Manager	Maintenance Manager	Maintenance Lead	Technician	Maintenance Planner	Maintenance Scheduler	Initiator	Reliability Engineer	Maintenance Buyer	Stockroom Clerk	Contractor
Step Number	**Step Description**																		
WCO-010	Work Execution																		
WCO-020	Review all completed W/Os									A				R					
WCO-030	Is the W/O properly documented?									A				R					
WCO-040	Return W/O for more info.									A		R		R					
WCO-050	TECO W/O									A				R					
	R = Responsible "The Doer"	Revision :	1.0																
	A = Accountable "The Buck Stops Here"																		
	C = Consult "Before a Decision"																		
	I = Inform "Keep in the Loop"																		

FIGURE 5.34 Work close out process RACI chart.

CHAPTER SIX

SUSTAINING AND IMPLEMENTING THE PROCESS

KEY PERFORMANCE INDICATORS AND REPORTING

Asset operating parameters, maintenance performance, and the maintenance work management processes themselves need to be monitored and adjusted to meet expected goals. Operations and maintenance work jointly in the best interest of the business to develop key performance indicators (KPIs) that are relevant to the assets and management processes that can be impacted by the frontline stakeholders who touch the assets on a daily basis.

I've been involved in too many situations where KPIs are set by stakeholders who are too far from the day-to-day execution of the processes and don't understand the effect of the selected KPIs on the organization. The need for KPIs should be understood, and the path to achieving positive results should be explained by the creator of the KPI. Too often KPIs are unobtainable, and this demoralizes the stakeholders, who believe "they" are being measured and failing.

The KPIs that are selected should be ones that the frontline stakeholders can affect and improve on with changes in performance, not behavior. I've seen

KPIs remain red for months, only to inquire about them and find that they were implemented because "it seemed like a good idea." After further analysis, it's been realized that the attribute being measured had no impact on reliability or performance.

At the outset of a new maintenance work management process design, start with a small number of KPIs, tracking the scheduled work and work execution. Consider the frontline stakeholders and how they can impact the KPIs in a positive way following the processes. Agree with the stakeholders on a path to improvement.

Set KPI compliance goals that reflect a reasonable level of improvement over today's actual performance, and tighten them over time. Setting industry best practice goals from the onset can hinder morale and performance. Meeting increasing goals with time creates an atmosphere of accomplishment.

Key Performance Indicator Criteria

KPIs are used and approved by management. However, management should work in collaboration with the frontline stakeholders and trust the frontline stakeholders to help set their own KPIs and goals. KPIs should be both performance relevant and process relevant. When working with the frontline stakeholders to develop the KPIs, ask the question, *Can the frontline stakeholders' performance impact the KPI in a positive way?* Ask, *Do the stakeholders know how they can improve their performance and that the KPI shows the results?* If not, question the KPI.

Consider KPI data collection mechanisms like computerized maintenance management system (CMMS) reports or other data mining software and how accessible and user-friendly the reporting is. Confirm the reliability of the data collection method. Make the data reporting easy to access and use. Dashboards are a quick, easily understood reflection of performance that can be displayed in easy-to-access areas by all stakeholders.

Maintenance Work Management Early Life Key Performance Indicators

At the beginning of your maintenance work management (MWM) design and implementation, start with a few performance and process-based KPIs; then expand in the future as agreed upon by management and the frontline stakeholders with a clear path to change performance in a positive way that leads to successfully achieving the desired performance levels. As performance goals are met, develop additional KPIs to improve and control other aspects of performance and process success only when needed. This is not an overnight progression. Consider the following to get started:

Schedule Break-in Work

Great effort should be taken to avoid schedule break-in work. Take the emotion out of decision-making by utilizing a formal risk analysis process that is approved by management. This typically reduces weekly schedule break-ins to less than 10%. KPIs should address:

- Percentage of work orders executed that are added to the locked schedule in a given time frame, typically the week 1 schedule
- Break-in work orders completed divided by total work orders completed × 100

Schedule Compliance

Planned and scheduled work costs are a fraction of the cost of emergency, unplanned, unscheduled work for too many reasons to explain here. With reduced schedule break-in work, more scheduled work is executed, and costs fall dramatically:

- Measures the overall effectiveness of the work management process as the percentage of craft hours on a schedule that are worked as scheduled.
- 100% of available craft hours are scheduled.
- Actual scheduled hours worked divided by total hours scheduled × 100.

Safety-Critical Work Orders Completed

Safety is the number-one concern in every organization. Safety-critical work orders should be the number-one priority on all schedules. Less than 100% is typically unacceptable:

- Safety-critical work orders completed divided by total safety critical work orders scheduled × 100

Preventive Maintenance Work Orders Completed

To protect the reliability of assets, preventive maintenance (PM) work is usually only second to safety-critical work, and high-achievement goals are expected:

- PM work orders completed divided by total PM work orders scheduled × 100

Planned Work

Well-thought-out planned work is safer, more efficient, and less costly, creating a relaxed execution atmosphere:

- Percentage of work orders on the schedule that are planned
- Planned work orders completed divided by total work orders scheduled × 100

Planned Versus Actual Labor

Work planning extends through to the highest levels of the business budget and impacts many aspects of the decision-making process. This KPI measures planning effectiveness, execution efficiency, and the adherence to the maintenance work management process:

- Percentage of actual craft hours performed on work orders completed versus planned craft hours for all scheduled work orders
- Actual craft hours to complete work divided by planned craft hours to complete all scheduled work orders × 100

CONTINUAL IMPROVEMENT FEEDBACK

Work Order Planning Feedback

Maintenance work planners are continually improving their skills and methods of planning. Planning feedback improves future plans and the associated safety benefits and cost savings. Planning feedback evaluates the success and improvement of the planning process, planning of labor estimates, work execution efficiencies, and material costs, leading to planner development and an ongoing improvement of the planning process. Many of the state-of-the-art CMMSs have functionality for both identifying and flagging work orders that contain planner feedback and providing fields for recording suggested improvements.

Work Order Reliability Engineering Feedback

To improve asset design, functionality, health, and reliability, reliability engineers depend on the experience and asset observations of maintenance technicians and equipment operators to provide real-time, day-to-day feedback on assets, using this information for improvements, replacements, and upgrades. This feedback is also used for PM optimization, equipment PM strategy improvements, and PM standard job plan improvements.

Engineering feedback is also used to evaluate the success of asset improvements and upgrades after they have been completed. As with work planning feedback, many of the state-of-the-art CMMSs have functionality for identifying and flagging work orders that contain engineering feedback and for providing fields for recording asset-specific behavior, asset condition, and suggested improvements.

Maintenance History

The common computerized maintenance management system is a very powerful tool, collecting data and storing standard job plans for doing work on assets and equipment. Standard job plans are used as templates when executing corrective work and satisfying equipment strategies defining preventive maintenance and predictive maintenance (PM/PdM) work execution.

The CMMS also stores all maintenance work data, which is invaluable to the maintenance planners, reliability engineers, project engineers, accounting, supply chain, inventory, maintenance and operations supervision, and management and regulatory. In industry, regulatory is widely known as government regulation, and is used for the entire life cycle of all assets.

Job plan libraries are used to calculate the cost of maintaining an asset or group of assets, developing the maintenance budget, determining the effectiveness of the preventive and predictive maintenance processes, identifying bad actor assets or equipment classes, and evaluating the condition of future investment in assets.

Maintenance work order history is used for creating new work notifications, looking for existing job plans and packages for needed repairs, finding job plans and packages for the repair of similar assets, and finding job plans and packages that can be modified for similar repair on a different asset, saving the planners valuable time writing new plans.

Maintenance cost history is used to determine if replacement or repair is a better option, to review frequency of inspections and repairs to determine if the PM and PdM strategies need adjustment, and to support PM optimization through reliability-centered maintenance and failure mode and effects analysis.

MAINTENANCE STEWARDSHIP

As part of the strategic continual improvement process, maintenance and operations meet on a regular basis to discuss maintenance work management process effectiveness and actual maintenance performance to the processes, agreeing on plans for improvements and making adjustments through specific actions and responsibilities. The need for evaluation and improvement is reflected in the established KPIs.

In most cases, operations and maintenance management also review monthly budgets, performance to the budgets, and needed adjustments to satisfy the budgets during the maintenance stewardship meeting.

Critical Elements

Elements include:

- **MAINTENANCE STEWARDSHIP MEETING AGENDA.** A regularly scheduled management review and improvement meeting (Figure 6.1)
- **MAINTENANCE STEWARDSHIP MEETING SCHEDULE.** Typically monthly to coincide with monthly budget tracking and reporting (Figure 6.2, shown later in this chapter)

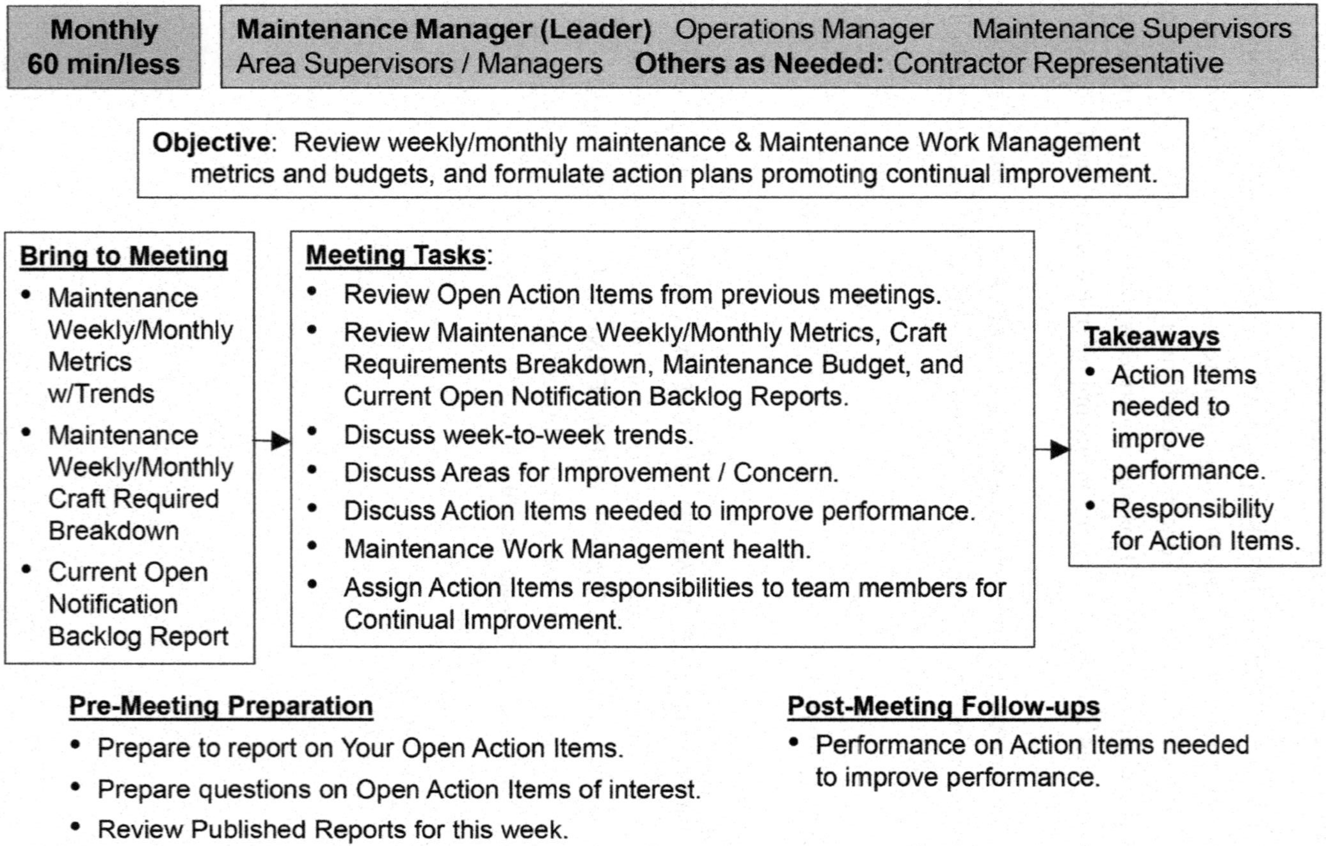

FIGURE 6.1 Maintenance stewardship meeting agenda.

MAINTENANCE WORK MANAGEMENT PROCESS IMPLEMENTATION

The implementation of the MWM processes varies widely, depending on the organization's culture and investment in documentation. My first MWM project was for an extremely well-disciplined top-five oil, gas, and chemical downstream organization. From the second level of supervision to the top of the operations, maintenance, and reliability departments, almost all positions were staffed with seasoned engineers.

This organization wanted only a wall-sized flowchart with swim lanes and an integrated RACI chart tying each of the six MWM processes together as one continuous flowchart. This flowchart was 3 feet tall and 11 feet wide, which fit the height parameters of the plotter in the engineering department. We embedded all the client-specific graphics similar to the ones used in this book on the flowchart at the point in the one big process where they were utilized. The one big RACI chart was at the top center of this graphical training and advertising tool. After implementation, a copy of this flowchart was hung prominently in each functional department.

A second-level operations supervisor and a second-level maintenance supervisor led a team with operations and maintenance supervisors, maintenance planners, maintenance schedulers, supply chain personnel, and a CMMS administrator to design the process. The second-level operations and maintenance supervisors took full responsibility for the design, implementation, and sustainability of the process. The second-level operations supervisor and second-level maintenance supervisor personally rolled out the process at the various plant area locations as a team to all cross-functional stakeholders, taking one hour to step through the processes and the roles and responsibilities reflected in the RACI chart and swim lanes, while answering clarifying questions.

With this organization, we also designed an audit to be conducted yearly. Their project charter ensures that any future changes to the process have to be submitted to, researched by, and approved by a committee made up of the original design team members/MWM subject-matter experts (SMEs), who were replaced in time if they moved to new roles. This model was repeated at five other locations over a six-year period. Subsequently, an external audit I conducted at one of the locations after a five-year period produced an overall rating to the original audit process of over 75%. The wall-sized flowcharts were still in place with one revision made after an internal request and review.

At the other end of the spectrum, I've been involved in highly documented process environments where the large flowchart was expanded by the addition of six separate MWM process flowcharts, step definitions, RACI charts, and a detailed maintenance work management manual supported by PowerPoint presentations that we used for training both online and offline.

In another organization, also a very well-known oil and gas company, we helped develop and deploy an automated self-auditing process that contacted all responsible and accountable stakeholders per the RACI charts automatically at

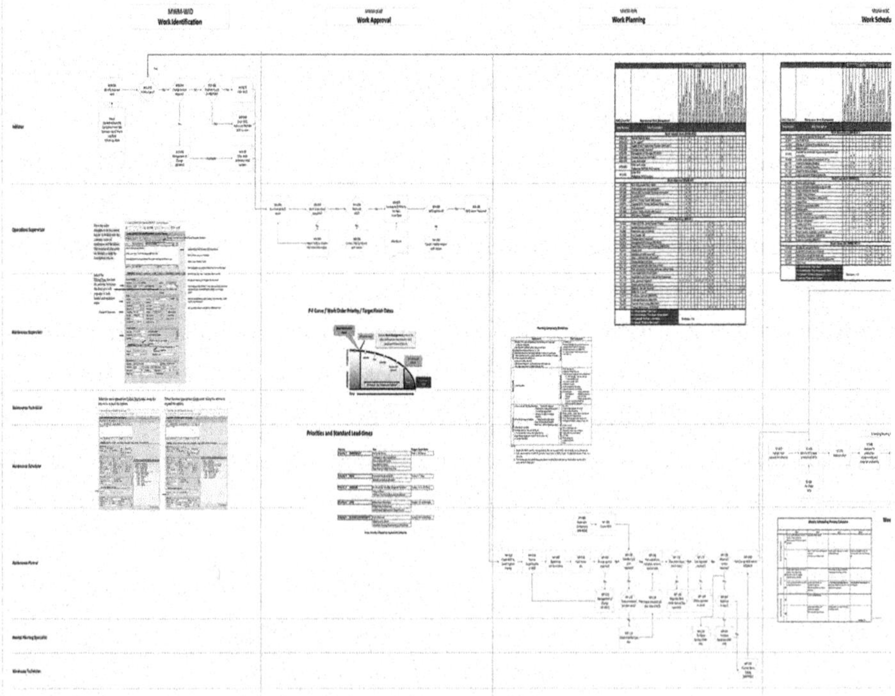

FIGURE 6.2 Integrated maintenance work management wall-sized flowchart.

various intervals, asking them to audit their performance using a short electronic form with yes and no questions that feed results to a database that reported the anonymous results. Our follow-up third-party audits showed a high correlation to the self-audits. The selection of your sustainability efforts will be tailored to your organization, culture, and budget.

Implementation/Rollout Presentation

In all cases, I have found that the inclusion of the one integrated process wall-sized flowchart works best for training all stakeholders as a cross-functional group in one location emphasizing the links among the different functions and their individual roles (Figure 6.2). I am accustomed to hanging the flowchart for future

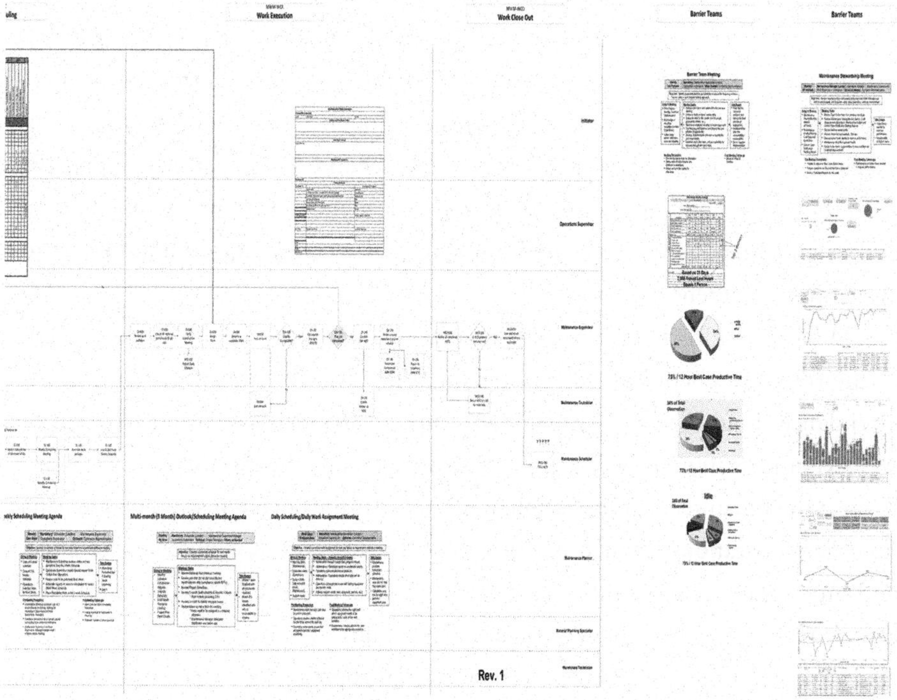

FIGURE 6.2 Integrated maintenance work management wall-sized flowchart (*continued*).

reference in all functional areas, especially all maintenance, operations, and reliability areas, as the whole organization tends to rely on maintenance stakeholders for much more than their intended purpose, and the process reminds everyone of the roles and responsibilities of the MWM stakeholders.

Second-Line Supervisors as Champions and Rollout Presenters

As discussed previously, the second-line operations and maintenance supervisors are ideal rollout presenters as they are viewed as the department managers. And the stakeholders who are responsible for the day-to-day MWM process execution are their direct reports. This is reflected in the process RACIs with the second line operations and maintenance supervisors being accountable for those responsible for most process steps.

Utilize the design team member operations second-line supervisor for the rollout of work identification and approval, and use the design team member maintenance second-line supervisor for the rollout of work planning, scheduling, execution, and close out, with the operations second-line supervisor supporting with the work scheduling rollout. As rollout leaders, the second-line operations and second-line maintenance supervisors reinforce their commitment to the process and its adherence.

When planning for the MWM process rollout sessions, publish and communicate training schedules and locations well in advance to ensure attendance. In attendance should be cross-functional groups by areas that are responsible for the MWM process steps, as well as all areas that depend on the execution of the process to meet their maintenance execution needs. Include the operations and maintenance managers, operations and maintenance frontline supervisors, maintenance planners and supervisors, maintenance schedulers and supervisors, CMMS administrators, inventory and supply chain personnel, and reliability-focused engineers.

When planning, ensure that you choose an appropriately sized training facility to accommodate all attendees, and be sure that the facility is designed with audio and video support equipment to communicate clearly. Provide ample time for questions and answers. Make a record of positive feedback and recommended improvements for future consideration. Record and follow up on areas where further clarification of the MWM process functionality and responsibilities are needed.

Implementing the contents of this book doesn't happen overnight. In small organizations, expect to spend four to six months, and in medium-to-large organizations, expect to spend eight to twelve months to achieve sustainability. Be patient with your team and all stakeholders.

MAINTENANCE WORK MANAGEMENT QUICK WINS

As the name suggests, there are benefits to be gained from the beginning of the MWM transformation project journey. The design process begins with the first of the six MWM processes—the work identification process—and progresses along the same path as the flow of the maintenance notifications and work orders.

As you design, validate, and approve each of the six processes, you can implement some of the changes to produce immediate benefits prior to the complete process rollout and implementation.

Maintenance Notification Quality

Due to poor-quality notifications, many hours are lost by downstream stakeholders who have to engage in verbal and written communications to clarify information that the work initiators and screeners should have provided in the notifications and work orders. This consumes many hours that the downstream stakeholders could be using to do their jobs providing the best product they can to their downstream stakeholders.

Risk-Based Work Selection and Schedule Break-in Work

Without fail, in my experience, work perceived as high priority or emergency can be greatly reduced by setting realistic, risk-based priorities and target finish dates. When risk-based priorities and target finish dates are applied using emotion-driven, knee-jerk reactions, the whole MWM value train is disrupted.

Typically, when I have begun the MWM transformation journey, a lack of organized, well-defined maintenance work planning and scheduling exists, and the usual recorded and actual schedule break-in rate is between 35% and 45%, causing many organizations to wonder why they bother to plan and schedule at all. The losses associated with their time-on-tools (ToT) study shows up in real dollars lost and the need for structured risk-based work selection, planning, and scheduling.

When carefully screening new work and setting risk-based priorities and target finish dates, weekly schedule break-in work drops below 10%, allowing ToT to increase as technicians are not interrupted while executing the planned and scheduled work.

Daily Scheduling/Daily Work Assignment

When starting to use the daily work assignment tool the afternoon before execution, maintenance frontline supervisors can build the daily schedule using their knowledge to set task duration.

When using this knowledge and assigning enough work to fill the technicians' day, the supervisors motivate the technicians to meet expectations. Coupled with the decrease in schedule break-in work, the amount of work completed each shift increases dramatically.

Execution Barrier Tracking and Elimination

Barrier tracking and elimination on a daily basis quickly resolve high-frequency barriers to work completion. This effort is compounded when involving other stakeholder groups like operations, supply chain, logistics, and contractors.

Maintenance Work Management Quick Win Benefits

When considering the cost of the MWM transformation effort, the sooner the organization can realize and quantify savings to help pay for the project investment, the better. In my experience, the maintenance work management transformation project becomes self-funding after 16–24 weeks, and the data collection process is easily used to calculate ongoing yearly savings.

MAINTENANCE WORK MANAGEMENT PROCESS SUSTAINABILITY

Often overlooked and underappreciated during the design and implementation of the new maintenance work management process is the ability of the process to sustain over time as organizations evolve and stakeholders come and go. A well-designed and well-implemented process needs to be protected from the future preferences of new stakeholders who are not process-oriented and most times see maintenance as a cost and not an investment. In maintenance and reliability, the low-cost option almost always deteriorates asset health.

Employee Development

The MWM processes and performance to the processes fit well into the typical employee training and review process. Within existing job descriptions, assign training and evaluation assignments by job function from the MWM processes' RACI charts. Verify that role-specific training has taken place, and evaluate MWM processes' responsibility adherence as part of the employee review process.

Periodic Maintenance Work Management Audits

The MWM scorecard used during the analysis phase of the MWM project is tailored to the details of your organization's six MWM processes. Using the same structure and numerical scoring system, you ensure consistency of evaluation over time, adhering to one evaluation and auditing process. Using a third-party, independent auditor can ensure impartiality.

As part of the sustainability plan for your MWM processes, schedule periodic MWM audits to review the processes and recommend improvements. In my experience, these are typically scheduled at 6-month intervals for a couple of years after MWM process implementation, then shifted to yearly.

CASE STUDY

MWM SUSTAINABILITY RESULTS CASE STUDY

Seven years after completing a 52-week MWM processes design and implementation with a team of 16 consultants for a very large oil and gas client with a $1.4 billion maintenance budget, I conducted a follow-up audit with one other consultant. The client had done a great job of sustaining their processes and had conducted a full review of all the MWM processes after a two-year period and upgraded their processes with one major improvement. We performed the audit over a four-week period, attending all MWM meetings and conducting interviews and Day In the Life Ofs (DILOs).

Here are some of our findings:

Positive Recognition

- All interviewees were very cooperative and transparent.
- The auditors were very helpful with coordination and introductions.
- The process flowchart was displayed in the production coordinator's area.
- Best practices do exist.
- Barrier capture, roll-up, resolution teams, and meetings are being utilized.
- Maintenance second-line supervisors are engaged with the MWM processes.

Meetings, Interviews, and DILOs Attended

These include:

Meetings Attended (30)

- Stewardship meeting (3)
- Work screening meeting (4)
- Schedule development meeting (2)
- Schedule acceptance meeting (1)
- Multi-month scheduling meeting (0)
- Daily update meeting (9)
- PMT meeting (3)
- Weekly planner meeting (0)
- Shift turnover meeting (4)
- Daily work assignment (4)

Interviews (35)/DILOs (4)

- Maintenance SLSs (3)
- Business team leaders (3)
- Production coordinators (6)
- Planners (6)
- Schedulers (1)
- FLSs (6)
- Production (10)

Backlog Management

The client was experiencing lower-than-acceptable schedule compliance and higher-than-acceptable schedule break-in rates. They developed a data analysis tool and were able to easily produce the schedule compliance report shown in Figure 6.3. Based on the client's low schedule compliance and excessive schedule break-in rates, they were confident that they would score a less than desirable ToT.

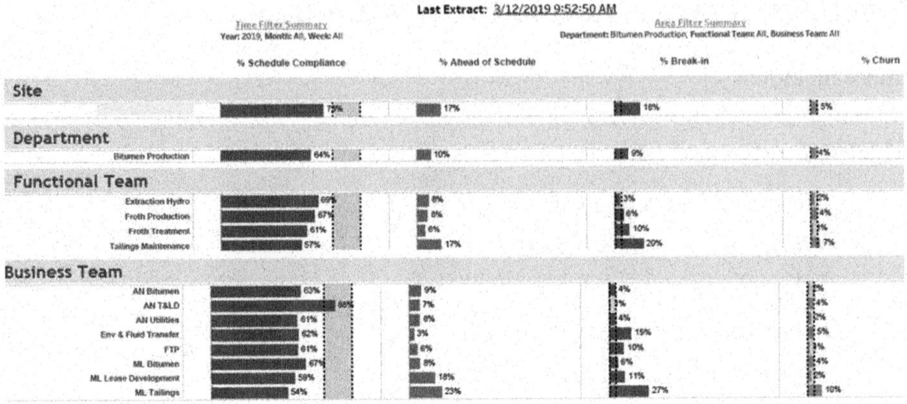

FIGURE 6.3 Schedule compliance.

Figures 6.4–6.8 show individual scorecards for each of the work management processes—identification, work approval, work planning, work scheduling, work execution, and work close out—along with the scorecard for barrier team and maintenance stewardship. Figure 6.9 sums up the information for the scorecards.

Maintenance Work Management Audit Scorecard

Wednesday, February 27, 2019

Criteria	Pts	Score	Comments
Work Identification and Approval			
Is the work selection process documented? Is there clear understanding of the GRS 6D workflow? a. Is there a posted work flowchart? b. Have all FLSs been trained to the GRS roles & responsibilities?	10	7.5	Found 2 process flowcharts displayed, found many copies of subprocesses with stakeholders, not all FLSs have been trained. Could use refresher.
Work Screening – Have the "Perfect Work Order" criteria been communicated to the business team Process Operators (or other work order quality criteria)?	10	8	Observed Work Screening Meetings. Found no excessive need for clarification from work initiators. Minor need for clarification was observed.
Are work notification descriptions complete -- Review sample of 20 work orders that have been planned within the last 60 days. -- Review a current sample of work notifications.	10	9	Descriptions were more than adequate for understanding the defect observed.
Are Work Notifications being screened daily? Quality & Completeness.	10	10	Observed 5 work screening meetings. The meetings are occurring daily, and a minimal backlog is maintained.
Work Screening – Meeting: a. Are work screening meetings being held daily? b. Is there a standing agenda, and is it being followed? c. Is the meeting duration 15 minutes or less?	10	7.5	Yes to a, c, d. Standing agenda is incorrect and needs updating.
Are Hurdle Rates & Break-in criteria documented? Have Hurdle Rates & Break-in criteria been communicated to the business team? Confirm by demonstrated examples.	10	10	Screening meeting members discussing each notification and applying established criteria. Team making every effort to apply break-in and hurdle rates providing Execution with ample time to complete with a realistic LACD.
Does break-in work exceed 5% of work orders completed? If more then 5%, what is the approx. % (> 10% = 5 pts, > 20% = 0 pts)	10	0	Unable to determine. Break-in rates reported in Stewardship Meeting do not reflect the information provided by the FLS on the schedule feedback form. Sometimes using LACD to set priority rather than using SAP status to drive Execution of break-ins.
Does the Business Team have a BT break-in Stewardship meeting which includes Mechanical & Process representation (or are break-in reviews done to maintain compliance) ?	10	10	All break-ins are approved via standardized form with appropriate sign-off of stakeholders. Break-ins that are reported are reviewed in the Stewardship Meeting.
Have the work screeners completed Risk Assessment training and GRS 6D?	10	10	It does appear that the participants have a high level of understanding of the work screening process, including Risk Assessment.
Are Service Deliverable Dates (SDD) being using as the basis for selecting LACDs for all work orders? Review sample of 20 work orders.	10	10	PCs are aware of the SDD and using SDDs to set the LACD.
	100	82	

FIGURE 6.4 MWM audit scorecard for work identification and approval.

Maintenance Work Management Audit Scorecard

Wednesday, February 27, 2019

Criteria	Pts	Score	Comments
Work Planning			
Is Material Coordinator in place and functioning full-time?	10	0	Due to the use of contractor staff and facilities, an ad hoc staff of Material Planners and Materials Handlers has been put in place. This staff is assisted and directed by the planners. Planners, FLSs, and PC are chasing parts.
Are jobs being "stepped out" (detail tasks/subtasks) with estimated times (or unit rate) for both XOM and Contractor crafts? Review sample of 20 work orders.	10	10	Yes, reviewed 15 work orders with task lists and bills of materials. Planners reviewed plan details explaining the execution of each job.
Is the Planner removed from the day-to-day execution of work - Emergency jobs and expediting material?	10	6	Planners are involved with the day-to-day expediting of materials when issues are discovered in execution. Planners are involved with break-in planning helping the FLSs. FLSs were discussing the lack of planning support on the weekend.
Does the Planner attend the weekly planning meeting (multi-craft jobs planning) each week?	10	10	Yes, the weekly planning meeting is utilized for complex planning as well as departmental and procedural improvements.
Are Planners field-scoping jobs and utilizing a compliant planning checklist?	10	5	The level of planning and feedback from the stakeholders indicates that most planners are field-scoping jobs. FLSs indicate that field visits do not occur with all planners.
Are parts and material identified on the job plan?	10	10	Yes, reviewed 15 work orders which contained comprehensive Bills of Material. These work orders were used for the planning of Preventive Maintenance, Functional Failure, and Chart of Accounts (Outage) work.
Are the Simple, Basic, and Complex job plan criteria documented and utilized? Is the 3-month "look-ahead" schedule being used?	10	10	Job Complexity Guidelines are posted in the area, and the status of planned jobs is communicated during long-term planning meetings.
Is there stewardship in place to track planners' effectiveness (or planners' scorecard utilized)?	10	5	No planner scorecard was observed. Plans are reviewed to determine planner competency and training needs. Planner backlogs are examined in the GRS 6D&E Stewardship Meeting. No clear separation between the Planning, Scheduling, and Execution Backlogs.
Does the Maintenance organization or Business Team have a feedback process on the effectiveness of job plans produced - accuracy of estimates, all required steps, appropriateness of parts and materials identified? Is there feedback to planning on plan quality (plan vs. actual)?	10	10	Yes, Planners and FLSs work closely during the development of plans. Any planning gaps observed during execution are communicated using a planning feedback form.
Are job packs being developed and properly utilized?	10	10	Reviewed 10 work packs. All work packs were complete per the Planning Complexity Guidelines.
	100	76	

FIGURE 6.5 MWM audit scorecard for work planning.

Maintenance Work Management Audit Scorecard			
			Wednesday, February 27, 2019
Criteria	**Pts**	**Score**	**Comments**
Work Scheduling			
Weekly Scheduling -- Are the Weekly Schedule Develop and Acceptance Meeting Agendas Followed? a. Review status of any deviations from this week's schedule b. Review work to be performed next week c. Assure Planning status is "Ready to Schedule" d. Determine when to reschedule remaining work	20	20	Yes, both meeting agendas are followed. All participants are prepared with revisions. Workload conflicts are resolved during the meetings.
Weekly Scheduling Meeting -- Are required attendees present? a. Mechanical Scheduler b. Process Unit FLS c. Mechanical FLS d. Process SLS & Mechanical SLS (optional)	20	20	All required attend with high level of participation.
Multi-Month Scheduling -- Are meeting attendees prepared for the meeting with the following: a. Copy of 4-week schedule b. Copy of current week's schedule c. Knowledge of job planning status d. Process - discuss and justify exceptions (only) to the 4-week schedule	10	5	Future time periods are represented with hard-copy schedules used by the PCs to convey changes. A Workforce Availability Tool has been developed that pulls all work order data from ISTe and allows the meeting participants to examine future time periods and make adjustments agreed to by all teams. A formal Multi-Month Scheduling Meeting as defined is not being held.
Weekly Scheduling -- Is the weekly schedule (frozen) completed by 3:00pm on Thursday for the next workweek?	10	10	Yes
Daily Update Meeting -- Is the Daily Update Meeting (DAS Meeting) Agenda followed? a. Review progress of today's DAS and identify jobs to carry over to next day b. Identify jobs/tasks that break the Weekly Schedule but need to be scheduled for the next workday c. Identify "first start" jobs/tasks that require preparation by Process and notify affected Units d. Discuss barriers/delays encountered during the current workday and identify possible corrective action	20	10	Maintenance SLS not chairing this meeting. The meeting follows the agenda; however, the DAS has been abandoned. In some meetings, work assignments are communicated using the whiteboard. The DAS Barrier aspect has been replaced by an abbreviated process whose effectiveness is less than adequate.
Daily Scheduling -- Is the Daily Assignment Sheet (DAS) completed and posted for the next workday so craft personnel can review it?	10	0	No, hand-written list on a whiteboard in one area. In other area lead craft handed out the work orders to be worked that day in the morning work assignment meeting.
Are all base maintenance resources (Syncrude / contractor) scheduled weekly? (< 50% = 0 pts, < 75% = 5 pts)	10	10	All resources, by craft classification, are loaded. However, resources are overloaded to create contingency work to compensate for shifting equipment availability. A lack of reliability between trains creates a low level of schedule attainment, as production demands for each train change daily. Work within the week 1 schedule is often delayed as workflow is shifted from train to train to compensate for equipment outages.
	100	75	

FIGURE 6.6 MWM audit scorecard for work scheduling.

Maintenance Work Management Audit Scorecard

Wednesday, February 27, 2019

Criteria	Pts	Score	Comments
Work Execution and Close Out			
Does the Mechanical FLS use the DAS to make the job assignments in the morning, and are goals communicated indicating the amount of time a job/task is expected to take to complete?	10	0	No, the DAS tool and process is not used. No expected hours to complete tasks are communicated.
Is the Mechanical crew at the worksite with tools, parts, and materials ready to walk down the job by established start time?	10	5	Observed work delay due to the lack of parts on 2 of 4 starts.
Is Process aware of job preparation requirements, and has equipment been prepared for Mechanical to work on it? Are permits prepared in advance of the job?	10	7.5	Observed 4 starts, one had permit delays. No one was available to sign permit.
Are there any Permit delays? If yes, how long on average? Are permit delays being captured?	10	7.5	1-hour delay.
Does the Mechanical FLS spend time in the field (at least 50% of total time) conducting positive follow-up on the work in progress addressing barriers encountered and providing recognition to the crew for good performance? (>50% = 10 pts, > 30% = 5 pts, < 30 % = 0 pts.)	10	5	Other obligations and time constraints restricted the time that the FLSs spent in the field. When observed, they were averaging 30% to 40% in the field.
Are required parts and materials available, or do the crafts need to make a trip for additional parts? Is pre-work (e.g., scaffolding) scheduled and completed timely?	10	5	Observed work delay due to the lack of parts on 2 of 4 starts.
Is required transportation equipment (rigs, etc.) at the jobsite as scheduled?	10	10	Yes, equipment and operators were available when needed.
Is there obvious lost time observed, e.g., late starts, early quits, and long breaks? Are these delays being captured?	10	10	No personal delays were observed.
Has support craft work been coordinated so there is no lost time? (support craft scope included in the job plan and scheduled accordingly)	10	10	Yes, scaffolding and insulation are planned and scheduled and are meeting the schedule.
Are break-ins given directly to the execution FLS?	10	10	Yes, FLSs are fully involved with the break-ins as they occur. Complex break-in tasks get support from the planners.
	100	70	

FIGURE 6.7 MWM audit scorecard for work execution and work close out audit scorecard.

Maintenance Work Management Audit Scorecard			
			Wednesday, February 27, 2019
Criteria	**Pts**	**Score**	**Comments**
Barrier Team & Key Performance Indicators (KPIs)			
Was the last Barrier Team Meeting held within the last month?	10	0	No DAS mechanism exists for collecting and examining barrier data. Other work assignment options are being explored to replace the DAS. Other methods of barrier capture are being explored.
Is the Maintenance SLS the meeting leader?	10	0	Meeting not taking place.
Are "mandatory" team members present (Mechanical FLS, Mechanical Planner, Process FLS)?	10	0	Meeting not taking place.
Did the team follow the Agenda and use the problem solving process?	10	0	Meeting not taking place.
Were barriers resolved or an action plan developed to address barriers?	10	0	Meeting not taking place.
Were KPIs published this month?	10	10	Stewardship Meeting is held weekly on Tuesday. Performance metrics are reviewed and discussed for all areas.
Were the KPI's published weekly for the past month?	10	10	All statistics are maintained in Tableau and are accessible by all staff members.
Were craft input numbers and Forelog numbers published monthly for the past month?	10	10	Forelogs are examined and discussed in the Stewardship Meeting.
Were the Daily Assignment Sheet (DAS) KPIs published monthly for the past month?	10	0	DAS has been suspended.
Were the Planner KPIs published weekly for the past month?	10	10	Planning forelog is reported in the weekly Stewardship Meeting.
	100	40	

FIGURE 6.8 Barrier team and maintenance stewardship audit scorecard.

Maintenance Work Management Audit Scorecard			
			Wednesday, February 27, 2019
Work Identification and Approval	100	82	
Work Planning	100	76	
Work Scheduling	100	75	
Work Execution and Close Out	100	70	
Barrier Teams & Key Performance Indicators (KPIs)	100	40	
Total	**100**	**68.6**	

FIGURE 6.9 Maintenance work management audit scorecard.

Maintenance Work Management Audit Gaps

Audit gaps include the following:

Work Identification and Approval

- Operation team lead not reviewing notifications for quality
- Process flowchart posted in one of three areas
- Production coordinator with less than 5 years' experience not properly trained

- Long description template not being used for notification quality
- PCs not rejecting poor-quality notifications
- PCs speculating on intent of notifications
- Planners receiving work orders lacking proper information
- Planners researching work order definition
- Planners spending considerable time researching information needed

No Material Coordinators

- Contractor materials staff and facilities
- Contractor material planners and materials handlers
- Assisted and directed by the planners
- Planners, FLSs, and PC are chasing parts

Planners in Day-to-Day Execution/Materials Expediting

- Planners involved with day-to-day expediting of materials
- Planners involved with break-in planning
- FLSs discussing the lack of planning support on weekend

Multi-month look-ahead is not happening in accordance with expectation.

Daily Work Assignment and Barrier Capture Process Fragmented

- DWA not used in all areas
- DWA Microsoft Visual Basic form no longer supported by IT
- Data roll-up coming from DWA and backup Excel form
- DWA not used in required meetings

Maintenance Work Management Audit Recommendations

Recommendations include the following:

MWM Processes Review—RACI and Processes Need Updating

The last major updates to the MWM processes took place in 2014. Since then, the organization has changed and responsibilities have shifted—materials planner; meeting agenda attendees have changed, etc.

Time-on-Tools Business Case for Change—Dollarize Losses

Given the high break-in rate and shifting equipment availability, it can be assumed that ToT will be low and thus justify the cost of updating the processes and retraining all the stakeholders.

Training for All Stakeholders to New Revisions

It appears that all stakeholders who were in place in 2012 during the first revisions are well aware of the process and its application. Stakeholders with less than five years' experience were trained with OJT and have a less than adequate understanding of the processes.

Deep-Dive Planner Skills Assessment for Uniform Planning

Planners' backgrounds range from technician to engineer, and the planners appear to have had no formal planner training to a uniform standard.

Deep-Dive Planner Process Review and Training—Uniform Planning

A need exists to develop a planning standard with a training process to bring all planners to the same skill level.

Sort Planner Backlog

Work is being held back after it is planned and is not advanced to the ready-to-execute stage while waiting for materials. The planner backlog includes planned work that is waiting for materials. The planner is monitoring materials before advancing the work to scheduling. This creates clutter and extra work to determine the work to be planned. Some work doesn't get moved and is executed, requiring extra effort to sort and clean the planner backlog.

Disencumbering the Planners—Focus on Planning Only

The planners are outside of their swim lane helping in the screening and materials processes, creating a backlog containing work that is past the latest acceptable completion date.

Local Planner Network

Not all planners are on the same page or learning from each other's experiences. Planners across sites and business teams do not know each other and do not benefit from each other's learnings. They would all benefit from uniform engagement with each other.

Separation of Planning and Inventory

Skills sets for planners and inventory management are different, as indicated by a planning frontline supervisor, who does not have the skills to manage inventory. The skills for planning and inventory management are distinctly different.

Separation of Planning and Work Screening—Estimating Cheat Sheets for Production Coordinators

The production coordinators are depending on the planners to provide repair costs during the screening meetings, creating a percentage of the planning backlog that is past the latest acceptable completion date. Cheat sheets are utilized to set estimated average costs of different repair types to save time obtaining actual cost estimates by the planners for every future repair.

Sever Maintenance Frontline Supervisors' Dependence on Planners

The maintenance frontline supervisors have a daily dependence on the planners as materials coordinators and schedule break-in planners. The planners are planning break-ins and chasing parts. This is taking the planners away from their duties and creating a backlog with work past the latest acceptable completion dates. The maintenance frontline supervisors need planner training for advancing schedule break-ins.

Achieve Conformance Across All Teams—Variance Between Teams

Individual teams have different levels of understanding of the MWM processes. Training and retraining are required.

Develop Standard Daily Work Assignment and Barrier Capture Tool

The DWA and barrier processes are fragmented, and the DWA tool is broken. The talent exists to build a spreadsheet-based tool.

MAINTENANCE WORK MANAGEMENT—ACTUAL RESULTS

Over the past two decades, I have been involved in or led almost two dozen projects designing and implementing MWM for clients ranging from very large to small. Following is a sampling of the most successful results:

USA Refinery

Developed, implemented, and enhanced maintenance work management processes for onsite plant and equipment, which enabled the successful implementation of a CMMS system, thereby achieving $15 million per year in program savings with a consulting cost return on investment (ROI) of 10:1.

Canada, Onshore Upstream Gas Production

Developed and implemented onsite plant and equipment reliability and maintenance work management processes, identifying maintenance financial control process deficiencies and contributing to a monthly maintenance budget segregation and monitoring process designed to eliminate lag between cost realization and reporting.

Was responsible for process design and implementation, management and frontline training and coaching, and process effectiveness/measurement, with a focus on planning, resource loading, scheduling, execution, and reporting to meet newly created process KPIs and targeted returns on investment, creating $8 million in annual savings per year with ROI of 8:1.

USA Refinery

Led a team that developed, implemented, and enhanced onsite plant and equipment MWM processes, enabling the successful implementation of a CMMS, while identifying and saving $13 million per year in maintenance costs.

Global Onshore and Offshore Oil and Gas Production Facilities

Upgraded and implemented upstream and midstream oil and gas maintenance work management processes. Trained, evaluated, and coached operations team leads and managers, maintenance team leads, planners, executors, and schedulers, as well as materials managers and coordinators, utilizing world-class CMMS planning and scheduling tools. Trained, coached, and qualified local and regional trainer/coaches.

USA High-Tech Manufacturer

Developed and implemented corporatewide initiative for full maintenance and reliability program, including equipment hierarchy and PM strategies, maintenance work management, and operations tie-ins utilizing their CMMS and maintenance scheduling work management systems.

Personally facilitated the development of a maintenance work management process and documentation. Taking personal role with the maintenance manager to build a full-service maintenance management team, including a maintenance planner, scheduler, buyer, and stockroom attendant through advertising for, interviewing, selecting, and training three team members.

Oversaw the upgrading of planning, scheduling, backlog scrubbing, and alignment with CMMS convention feeding corporate-level KPI data collection.

Increased the standard work management scorecard from 17% to 84% through process development, training, and buy-in.

Canadian Refinery and Chemical Plant

Led team in developing, implementing, and enhancing onsite plant and equipment maintenance work management processes. Responsible for project management, process design and implementation, management and frontline training and coaching, process effectiveness, and measurement, with main focus on planning, resource loading, and daily schedule communications. Increased site time on tools from 36% to 52%, resulting in increased utilization of $3.876 million, and reduced contractor utilization from 225 to 195, resulting in decreased annual cost of $8.244 million.

American Refinery

Coached the operations leadership in adopting risk-based work selection and target finish date priority-driven planning, scheduling, execution, and engineering utilizing a well-known work planning CMMS, abandoning a homegrown database priority process, and enabling the implementation of a best-in-class scheduling CMMS software, while scrubbing operations and planning backlogs and reducing the break-in rate from 49% to 15%, identifying a shortage of work beginning in week 4 of the upcoming schedule.

American Gold and Copper Mining and Smelting Operation

Conducted analysis of reliability and maintenance work management processes and procedures, resulting in the forming and leading of an SME team to develop and implement detailed maintenance work management processes to increase the utilization of the company's CMMS and maintenance scheduling systems.

Worked closely with operations to improve CMMS work order quality while training and coaching risk-based work selection, enabling logical work order prioritization and the use of target finish dates for work center loading-based scheduling best practices.

Made key upgrades to planning and scheduling processes and employee skills through training and coaching, enabling task-level planning and scheduling. Taught a three-day planning/scheduling class to 44 planners, schedulers, and maintenance and operations supervisors.

American Onshore Upstream Gas Operations

During a period of downsizing, and implementing world-class CMMS planning and scheduling tools, working hand in hand with the planning and scheduling teams, provided individual skills assessments and coaching.

During this period, recognized with the planning scheduling manager that a gap existed with the definition of the roles and responsibilities under the new organizational structure and that the organization had no industry standard maintenance work management processes.

Formed a design team that included the operations, maintenance planning, scheduling, execution, and corporate maintenance work management teams to design processes for work identification, prioritization, planning, scheduling, execution, and closure/analysis. Provided implementation and training plan to be executed by the client.

EPILOGUE

The value and integration of processes is not new to me. At 17, I learned—while working for a worldwide manufacturing company—the interdependencies of all facets of a manufacturing operation. My curiosity to know and influence the big-picture formed the basis of my career. The next five decades saw me step through the entire process, holding positions in all departments except accounting and product engineering. I was highly dependent upon these functions, so I had to learn how they functioned and the depth of their value to my success.

I hope that with this book you can learn how to know where you are today. You have the knowledge and skills to take an objective look at the maintenance and reliability value train within your organization. You have the knowledge to identify the gaps that stand in the way of your organization to optimize their maintenance investment.

With this knowledge, I believe you can develop a roadmap and create a vision and plan to identify the state of knowing where you will be in the future. This is not an overnight process. There have been a few individuals in my last two decades of work who sought me out to let me know that they knew that my intent was cultural change. These individuals worked with me to sell the new way of managing.

In working with all stakeholders and considering and integrating their input, you will learn that your organization has the capacity for change. There will be hurdles to overcome. You will have to be the agent for change and sell your ideas. Even if you are the agent and sitting at the top of your organization, you will need advocates to reach every nook and cranny of the organization to make your effort successful.

You now have the knowledge to use the process science to effect positive change. This science isn't new. It has continuously developed since the mid-nineteenth century and started to become well-documented in the 1960s for manufacturing control. There are many details that you may have not understood before reading this book. It took me two decades of designing, implementing, and sustaining maintenance work management processes. This book will help you get there quicker.

It's important that you understand how you are helping your organization to reach its anticipated future state. It's necessary for you to include peers from all the functional areas of your organization to sell them on the value of the effort that you all will share. Your effort will need a team that can see the vision, believe in the vision, and do their part to realize the vision. No one is in this alone.

You must be the visionary, recognizing the interdependencies in your organization and the importance of all functions working towards one goal. The success of the organization depends on all stakeholders. Bringing them together will not be easy. They will have conflicting goals. You will have to be the objective mediator who shows that to optimize the process and execution, there will have to be compromise and then optimizing the compromise to create success at the organizational level.

This optimization will not live into the future on its own. It must be written into the processes and become a way of life for the entire organization and all stakeholders. Safeguards, like audits and editing responsibilities, supported by a written process and procedures, with all stakeholders trained and understanding their roles, and limited access to changing the process is vital. This is a science and not a personal preference for the untrained. It's a group effort with all stakeholders represented.

The contents of this book have helped me to create success at many levels for many organizations for my long and enjoyable career. I've had the opportunity to not only create success in these organizations, but have extended success to outside organizations with interdependencies to help them to optimize interactions. This has extended forward to customers as well as extending to the supplier base. It brought great satisfaction that you can also experience.

ACKNOWLEDGMENTS

I thank Denny D, Dennis Z, Bob, and the thousands of operations frontline managers, supervisors, and operators and maintenance frontline managers, supervisors, planners, schedulers, and technicians, as well as stockroom attendants, buyers, and reliability engineers who worked with me in the dirt doing our part and making a difference. I hold many memories that made this portion of my career a pleasure and fun. Thank you for being real and being a part of my experiences.

INDEX

Page numbers followed by *f* refer to figures.

ABOUT THE AUTHOR

DALE TANNER is a reliability, maintenance and operations management consultant and manager with extensive experience in systems/process analysis and development, process reengineering and improvement, work selection, planning, scheduling, execution, CMMS/CIM utilization, data analysis, supply chain, control of work, process mapping, project management, budgeting, and cost containment. He has 25 years of consulting experience including engagements with clients in the United States, Canada, South Africa, Papua New Guinea, and the United Kingdom in oil and gas, synthetic oil, gasification, chemical, oil-sand mining, up, mid and downstream gas and oil on and offshore production, coal mining, metals smelting and melting and discreet manufacturing industries including energy, consumer goods, communications, computer, food, OEM, government contract and aerospace industries. He has worked with such heavy-hitting companies as ExxonMobil, Imperial Oil, Chevron, Syncrude, Cenovus Energy, BP, Sasol Syncrude, Barrick Gold, Freeport-McMoRan, TimkenSteel, Stanley Tools, Digital Equipment, Fluor, and Redfield Rifle Scopes and has consulted through Strategic Asset Management Inc., The Marshall Institute, Mainnovation, The Highland Group, Life Cycle Engineering, Evolve Partners, Nexus Global, and Parkwood Consulting.